WHEN IT ALL BURNS

WHEN IT ALL BURNS

Fighting Fire in a Transformed World

Jordan Thomas

RIVERHEAD BOOKS
NEW YORK
2025

RIVERHEAD BOOKS
An imprint of Penguin Random House LLC
1745 Broadway, New York, NY 10019
penguinrandomhouse.com

Copyright © 2025 by Jordan Thomas
Penguin Random House values and supports copyright. Copyright fuels creativity, encourages diverse voices, promotes free speech, and creates a vibrant culture. Thank you for buying an authorized edition of this book and for complying with copyright laws by not reproducing, scanning, or distributing any part of it in any form without permission. You are supporting writers and allowing Penguin Random House to continue to publish books for every reader. Please note that no part of this book may be used or reproduced in any manner for the purpose of training artificial intelligence technologies or systems.

Riverhead and the R colophon are registered trademarks of Penguin Random House LLC.

Some of the material in Chapter 14 previously appeared
in different form in *The Drift* (Issue 2, October 2020).

Map by J. C. Franco

Book design by Daniel Lagin

LIBRARY OF CONGRESS CATALOGING-IN-PUBLICATION DATA
Names: Thomas, Jordan (Anthropologist), author.
Title: When it all burns : fighting fire in a transformed world / Jordan Thomas.
Description: New York : Riverhead Books, 2025.
Identifiers: LCCN 2024034346 (print) | LCCN 2024034347 (ebook) |
ISBN 9780593544822 (hardcover) | ISBN 9780593544846 (ebook)
Subjects: LCSH: Wildfires—California—Prevention and control. |
Wildfires—Climatic factors—California.
Classification: LCC SD421.3 .T46 2025 (print) | LCC SD421.3 (ebook) |
DDC 634.9/618—dc23/eng/20240730
LC record available at https://lccn.loc.gov/2024034346
LC ebook record available at https://lccn.loc.gov/2024034347

Printed in the United States of America
1st Printing

Some names and identifying characteristics have been changed
to protect the privacy of the individuals involved.

The authorized representative in the EU for product safety and compliance
is Penguin Random House Ireland, Morrison Chambers, 32 Nassau Street,
Dublin D02 YH68, Ireland, https://eu-contact.penguin.ie.

This book is for the hotshots.

For all the people on the front lines of the climate crisis.

And for Kathy Supple.

CONTENTS

Author's Note | ix

Introduction | 1

PART I: TRAINING
7

PART II: ATTACK
87

PART III: RUIN
159

PART IV: ESCALATION
199

PART V: REGENERATION
247

Afterword | 317

Acknowledgments | 321

A Note on Sources | 325

Notes | 327

AUTHOR'S NOTE

WHEN I BEGAN WRITING THIS BOOK, I WANTED TO WRITE about climate change and wildfires. I quickly realized, however, that it wouldn't be accurate to write about this topic without chronicling the historical factors that shaped the landscapes climate change now disrupts. This history includes the ways Indigenous people in California used—as many continue to use—fire to shape the land. It also includes the violent processes by and through which governments have attempted to take the use of fire away from Indigenous people through time.

As a white anthropologist and writer, this presented me with a dilemma. Any account of the megafires of today that does not centrally acknowledge Indigenous burning perpetuates an erasure of Indigenous people—including their histories and contemporary movements for cultural sovereignty. If a story of fire is to be told, it must not only include but also elevate the lived realities of Indigenous people and, ideally, be a voice from these diverse communities. This, however, is not me. And outsiders—often external anthropologists such as myself—have a disquieting history of narrating Indigenous stories on their behalf. This has often led to a failure to reflect people's lived realities and, at times, has perpetuated dominant systems that undermine Indigenous cultural sovereignty.

It is a fraught space in which to write. Honestly, I don't believe there is a clear resolution to this dilemma of erasure versus narration. Nor do I think there could or should be. Like many dilemmas, this one begs a constant revisiting, reflection, and attention.

In this book, I've done my best to draw from Indigenous sources when detailing their histories and current perspectives on fires. This includes using primary sources in the historical literature, interviewing Indigenous fire practitioners and activists working to revive cultural burning, and attempting to forefront the work of Indigenous scholars who provide critical perspectives on wildfire management.

The end result is far from perfect, I am sure, and I humbly request patience from my readers as I do my best to craft a story that is both accurate and ethical.

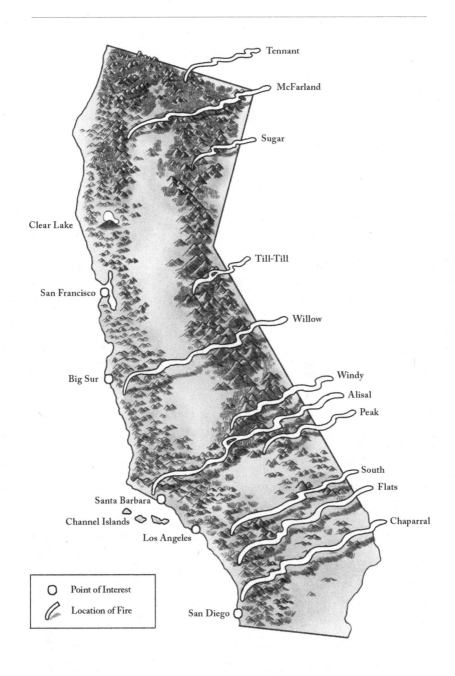

WHEN IT ALL BURNS

INTRODUCTION

ONE AFTERNOON IN SEPTEMBER 2021, AS A MEGAFIRE BURNED through Sequoia National Forest, my hotshot crew marched into a grove of its ancient trees. In the crisp air of high country, the sounds of creaking wood and swishing pine muted our footsteps. Sequoias towered through the smoke. Their canopy closed hundreds of feet above. We moved between trunks the size of cabins, the bark grooved and red. Small squads dropped off from our crew of twenty, spreading along the ridge to prepare for the megafire hurtling toward us.

My own group peeled off near the top of the ridge to await orders. After fighting this fire for the past week, my knees ached, my blisters stung, and my head hurt. I was tired. I knew the other hotshots felt the same. They were chewing tobacco and sharpening their tools—hybridized shovels, axes, and hoes with macho names: the Rogue, Pig, Chingadera. I set my chainsaw on the forest floor, removed my gloves, and, for the fifth time that day, pricked my thumb on the chain's teeth to make sure they were sharp. The grove made me restless.

In the past two decades, wildfires have been doing things not even computer models can predict, environmental events that have scientists racking their brains for appropriately dystopian terminology: firenadoes, firestorms, gigafires, megafires. Scientists recently invented the term "megafire" to describe wildfires that behave in ways that would have been impossible just a generation ago, burning through winter, exploding in the night, and devastating landscapes historically impervious to incendiary destruction—like the sequoia groves of California. Sequoias are among the oldest organisms in existence, with fire-resistant bark several feet thick and crowns that can recover when 90 percent is scorched. They even rely on fire to reproduce, as flames crack their cones so seedlings can germinate. Now, the same ecological force they once depended upon is pushing them toward extinction.

Sequoias' lives are monuments of deep time. Their death would signify something else. If we could not hold this ridge against the megafire, the sequoias would become the largest torches on earth, carrying flames higher than the Statue of Liberty. After three thousand years of life, they would become charred monuments to a passing era, symbols of a violent future. By holding the ridge, I felt we were holding back a new, altered world. As I waited on the fireline with ash in my lungs, I still hoped this was possible.

A figure appeared through the smoke, huffing toward us. I stood, worried our superintendent, Aoki, the leader of the Los Padres Hotshots, would catch us sitting. I relaxed when I recognized Jackson.

"Jack!" someone called.

"What's the word?" another asked.

"Fellas," Jack greeted us, loosening his pack to catch his breath. His face and beard were caked in dirt. "Cancel your plans, boys. We're gonna be up all night." Word had traveled down the chain of command that we were going to burn our line that evening.

We had spent six days building that line, working from sunrise to sunset, cutting a path through brush and hills, up a mountain ridge into high timber and finally to an old logging road. Wildfires cannot burn without fuel—grass, bushes, and trees—and our line formed a fuel break, a continuous barrier of dirt that snaked around the megafire. When the conditions were right, we would set fire to its edges. The flames we lit would move toward the megafire and consume the fuel in its path. This is the fireline—a dirt band that holds fire back from the world.

Now, we only had a few hours before the megafire would hit us—a few hours to finish the line and set fire to the ridge.

"It's gonna be a shitty burn," Jack relayed. "We'll be chasing spot fires all night."

I asked if we had any safety zones, defensible spaces like meadows where we could survive if the fire got out of control.

Jack hedged, chewing the corner of his mustache. "There're some bulldozers trying to plow one near the peak," he said. "The thing is, we'll be going downhill. So, if this thing gets away from us, we'll need to run to our vehicles and get out of here."

"It's a forty-five-minute drive out of the forest," I noted.

"Yeah, hopefully we'll get out."

A radio crackled. Through the static came a gruff voice. Márlon was reporting from his lookout position. "The fire seems a *little* more extreme," he said. "Three-hundred-foot flames coming at us."

Aoki's laughter cut into the radio traffic. "Hoookay! Let's get the rest of the line pushed through and get that burn goin'."

My group huddled in the trees in smoke and radio silence. We didn't know where Aoki was or who he was telling to push the line. We often joked that he was a mountain spirit. Tall and willow thin, with black hair that fell to the small of his back, he moved with the smoke, drifting and reappearing through the trees. Aoki had turned fifty that season and was widely considered the most experienced hotshot in California—a state that claims the most skilled firefighters globally.

Now, the megafire was throwing embers ahead of its advance, starting smaller fires, spot fires, like raiding parties for an approaching army. Axel, a squad boss, called us toward one that was smoking in the valley below. If we let it grow, it would rush up and cross our line, igniting the forest behind us and trapping us in the path of the megafire. "If the fire gets behind us," Axel said, "we're fucked."

Scrambling down loose rocks into the valley, we followed Axel's shout and found the spot fire. It had grown to the size of a football field. We cut around it, choked with smoke. In the ringing silence that follows a chainsaw's scream, I realized that the silence was becoming a roar, and the roar came from the megafire. It was close.

The dark understory of the forest began to glow, a vibration emanating from its depths, rising to the sound of a jet engine. The treetops groaned with wind created by the force of the megafire.

On the fireline, I rarely felt adrenaline. I experienced danger more as a pressure, a weight that never fully disappeared. As the danger grew, so did the weight. With the megafire roaring through the trees, I'd dropped my chainsaw and begun filing the saw's teeth to new points. Now, the chainsaw shook in my hands.

"Listen." Axel cocked his head with wild eyes and a feral grin. "You hear it? *Let's get out of here.*"

Axel led our escape, men sweating as the moon rose like a counterweight to the setting sun. A radio crackled, Aoki warning that the fireline was no longer safe and we should hurry down the ridge to our vehicles. I shouted to Axel, several headlamps ahead, that we were throwing away the past week of work. "It's part of the job," Axel replied. "You'll get used to it." His reply evoked a hotshot adage: *You learn to let go of hope, or you get crushed.*

As we walked, our headlamps cast beams through clouds of smoke and dust, illuminating the trees around us like pillars supporting another world. Forced to retreat, I felt like we were abandoning that world. Just before reaching our vehicles, I stopped to look back at the megafire. It was a billowing column of shadow and light. Briefly, I doubted what I'd seen: through a gap in the trees, a lone flame rose and flickered above the rest.

Only a sequoia—a tree that had coexisted with fire for millennia—could carry flames so high.

Part I

TRAINING

CHAPTER 1

One Year Earlier

IN AUGUST 2020, SOON AFTER I FIRST SIGNED UP TO SUPPRESS WILD-fires for the United States Forest Service, a puff of white smoke was reported in Big Sur, California. The blacktop strip of Highway 1 cut along coastal cliffs, speckled with vehicles that seemed to dangle over the ocean. Helicopters approached the ocean in an insect procession, scooping water in big balloons, which they then dumped into a flaming gorge. Within that gorge, clusters of homes, with their carefully tended orange trees and gardens, were evacuated. Redwood trees creaked and groaned in cloaks of moss, hidden in the shadows of late afternoon. Heat distorted the air until the whole land seemed to flutter like windblown curtains.

I stood on a ridge with a young man who really wanted to look like a wildland firefighter. He had grown a mustache, bought a belt buckle, and smeared some soot on his cheeks, perhaps to hide the fact that, a week before, he had been selling kombucha in a beach town to the south. I had no more claim to the job than he did, having taken emergency leave from an anthropology graduate program at the University of California to fight this fire. We were both on a beginner crew, a patchwork of Trader Joe's cashiers, yogis, Zumba instructors, pet rescuers, painters, and fishermen, misfits of all stripes, all hoping to make a name for ourselves so we could rise in the ranks of the wildfire world. And we all stood in the thrall of the largest flames any of us had ever seen, watching them swallow a mountain across the valley.

"Incredible," said the young man beside me. His name was Finnegan. "All because one guy lit a match."

I looked at Finnegan sideways, annoyed, not because he spoke like he was on television, or because he was my rival—as a fellow "saw dog," he was gunning for my position as lead sawyer—but because I envied his simple cause-and-effect explanation for the unfolding destruction. On one hand, Finnegan's observation made sense: light a spark in the mountains, the mountains burn. On the other hand, the flames were, for me, much more puzzling.

I sought solidarity with another would-be firefighter. He was my age, late twenties, no mustache, a duck hunter who could name just about every plant in California. I pulled him aside. "Hey, man, have you, like, ever wondered what fire *is*?" Before the words left my mouth, I realized how stupid they sounded.

"I know what fire is."

"You do?"

"Yeah. It's the exothermic chemical process of rapid oxidation."

"Mmm."

I told myself I wasn't the only one struggling to understand wildfires. We were in the middle of the most catastrophic fire season in California's recorded history. The redwoods in the gorge below us were torching. The sequoias, their inland cousins, were burning almost to extinction. Four million acres of land would go up in smoke. The entire state was veiled in the red haze of perpetual dusk, and there didn't seem to be enough explanations to go around. The president at the time, Donald Trump, said we needed to rake our forests like the Scandinavians. Ryan Zinke, a fossil fuel pundit who had formerly been the top official charged with managing America's federal lands—which comprise around 47 percent of California's total acreage—passed the blame to environmentalists, claiming they weren't letting corporations log enough trees. Fox News said, "The fuel for California's wildfires is the government, not climate change." Meanwhile, as flames spread within three miles of Governor Gavin Newsom's commercial vineyards, the governor spoke against what he saw as attempts to shield the fossil fuel industry from complicity. "This is a climate damn emergency!" Newsom declared.

Did the fires come from climate change or land-use practices, governmental ineptitude or corporate malfeasance? Or maybe, just some guy who lit a match? The swirl of competing explanations reminded me of a common Buddhist parable in which a group of blind friends encounter an unfamiliar creature, an elephant, and begin touching it to discover what it is. One, with his hands on the trunk, believes it is a snake. The other, feeling the ear, calls it a fan. Another, leaning against a leg, believes it is a pillar. While none perceive the whole of the elephant, each believes that their subjective experience is absolute truth, and, suspecting that the others are deceiving them, the friends eventually come to blows. To stretch the metaphor, I felt like the wildfire was the elephant in the forest, and it was on a rampage.

The other firefighter's textbook recitation of oxidation and exothermic what-have-you went over my head, but he seemed to have a finger on the creature's pulse. Fire is unique to our planet because life is unique to our planet, and life, through a strange series of events, creates fire. Some five hundred million years ago, when plants first populated the land, they used the sun's energy to transform carbon and water into stems, leaves, and roots. Their waste product was oxygen, a highly reactive gas, which they exhaled into the atmosphere. When these same plants are heated by, say, lightning, their molecules break into vapor. That vapor reacts with oxygen to form fire. The flames that catch our eyes are simply particles of soot heated to incandescence.

Ever since the origins of terrestrial life, as forests have spread and shrunk and grown again, fires have moved in their midst, rising and falling and roaring and scurrying with the ebbs and flows of atmospheric oxygen and floral life until every surface has been touched by flames. Over these millennia, flora and flames developed an elegant choreography, a symbiosis of sorts, as some ecosystems grew to tolerate fire, others came to need fire, and some even contrived to produce the kinds of fires they need. Fire is as woven into the fabric of life on our planet as the rains that wash its hills and the rivers that flow through its valleys.

Yet humans have completely altered the forms fires take by changing the contexts in which they burn: the patterns of vegetation, frequency of ignitions, and, recently, carbon in the atmosphere. Each human society carries a unique environmental footprint. By extension, every society creates a unique fire regime. On that ridge in Big Sur, watching flames chew through redwoods and belch them into the sky in whirls of black smoke, I didn't have an easy answer to the question of what these flames were.

<hr>

UNLIKE MY NEW COLLEAGUES, MANY OF WHOM GREW UP WATCHING their fathers rush into the red, glowing night, I had only recently become interested in wildfires. Raised in the Midwest, far from America's explosive West Coast, I had always considered wildfires as a distant, separate problem. I found their portrayal in film and media boring—like a war story, but with less action and worse outfits. When I was young, wildfires rarely covered whole hemispheres with smoke.

It is easy to think of wildfires as a problem unique to the American West. Every year, news cycles follow the cycles of our planetary orbit. In the United States, flames appear in headlines as soon as the northern hemisphere tilts toward the sun. The air warms, the forests dry, and flames inevitably emerge. In recent years, these headlines have been embedded with warnings that our society is out of balance with nature, hurtling toward climate apocalypse. And these warnings are not entirely wrong. Eighteen of California's twenty largest wildfires on record have burned in the past two decades.

But I was approaching wildfires from a different angle. In my early twenties, I had spent seven months traveling from Kansas to South America, mostly on a bicycle. I wrote an essay about the experience, which landed me funding to research anthropology at the University of Cambridge, in England. I was skeptical of the idea that humans are inherently destructive to our environments, so for my research project, I returned to a Maya community in southern Mexico that had previously welcomed me. Indigenous people in this region have been sustainably managing forests for millennia. When I arrived in that village, I was surprised to see plumes of smoke rising over the jungle.

The Maya farmers, I learned, used fire to tend to the forest much as those in the region had for the past five thousand years. When the conditions were right—the wind was low, the humidity was high, and the temperature was cool—they lit fires that nourished the soil, encouraged useful plants, and enhanced the diversity of the forest. People in southern Mexico have been managing these forests ever since today's warm tropical conditions emerged from the last Ice Age, and their techniques have evolved with the trees. Without carefully tended flames, the Maya farmers told me, the health of their food systems, and of the forests, would be jeopardized.

While I was finishing my graduate degree amid the pointy buildings and dusty libraries of England, I read every book I could find on fire, trying to make sense of the fact that flaming landscapes aren't simply the stuff of breathless media commentaries and militaristic suppression campaigns, but can be integral ecological processes in landscapes as diverse as tropical jungles, African savannahs, arctic peatlands, and the California coast. Nearly every terrestrial area, I learned, has evolved with different kinds of fire—flames catered to each ecological niche and shaped by the people who inhabit the land. What did it mean, I wondered, that so much of the planet was now experiencing continental conflagrations?

This question was on my mind when the English summer skies became pale with smoke. Unprecedented wildfires were burning not just in California, but in Europe, North Africa, South Africa, Australia, Indonesia, the Arctic, and the Amazon. Historically, fire had a natural place in each of these ecosystems, but something had changed. The smoke wrapping itself around the earth was the effluence of infernos that erased forests, incinerated towns, killed hundreds of people, and drove species to the brink of extinction.

I wanted to make sense of the cultural currents that stoked this new era of fire, but my degree was nearing completion. My girlfriend, Kenzie, had recently quit her job at an English bakery and moved to California, where she was raised. I decided to follow her, but like most graduate students, I was broke. In a rustic attempt to preserve funds for the overpriced garage Kenzie had secured for our housing in Santa Barbara, California, I finished the edits to my thesis in a tent under a bridge beside the River Cam, listening to the patter of rain and drinking wine by the light of a headlamp. I was cold, damp, and far removed from the fires that captivated me. Still, it was romantic, I told myself, to spend my final weeks in Cambridge camped along the riverbank where Charles Darwin formulated his theory of evolution. "The discovery of fire," Darwin speculated, perhaps strolling along this very river, was "probably the greatest ever made by man, excepting language."

In 1871, the year Darwin published *The Descent of Man*, the notion that fire was man's greatest discovery was not original; it reflected the mythological structure of nearly every world culture. In Namibia, the Bergdama people tell of a man who stole fire from a pride of supernatural lions. For the Awikenoq of British Columbia, a deer spirit delivered fire as a gift to primal humans. In one Pueblo myth, the envoy was a coyote. The Greeks say that Prometheus filched fire from Mount Olympus and shared it with early humans. In each case, in each culture, in each story, fire is the bridge between our primal past and modern present, and its discovery illuminated the path toward civilization.

Myths are useful narrative tools, but they aren't known for their scientific accuracy. Like Christian creationism, the notion that fire was "discovered" condenses several million years, and a lot of complexity, into a single momentous event. In spite of this, over the years, the old myth of man and fire became woven into scientific language. Intellectuals who were instrumental in founding the academic disciplines of anthropology, psychology, evolutionary biology, and French philosophy all opined, at some point, about the pivotal role the discovery of fire played in the history of our species. Today, the idea of discovery enters popular consciousness most commonly in the form of cartoons, where a primitive man—always a man—invents fire; usually, the rhetorical implication is that men drive human progress through innovation.

Perhaps such cartoonish projections are enabled by the fact that the origin of humanity's relationship with fire remains quite mysterious. No one *really* knows exactly who used fire first, or where, when, or why they did so. This is because traces of fire are quickly washed away by rain. Of the evidence that does survive the wear of time, it is often impossible to distinguish between fires that burned naturally and those that were controlled by humans. Both lightning fires and firepits char animal bones. Smoldering roots harden clay just like hearths do. And many scientists, their vision constricted by their search for the man who brought light to humanity—the Elon Musk of cavemen—failed to imagine that our relationship with fire was perhaps not a discovery, invention, or conquest, but a gradual process that unfolded among the forebears of our species over several million years. However, this idea did occur to primatologist Jill Pruetz when flames began to crackle near the troop of chimpanzees that had adopted her in Senegal.

Pruetz heard the fire before she saw it, a rustling in the grass and a pop of saplings. Her anxiety began to mount when smoke blocked out the sun. But the chimpanzees around her in a small patch of forested ravine seemed undisturbed. As the flames jumped over the lip of the slope and crept downhill, an adult male woke from a nap and began grooming himself. When the fire reached the trees and leapt ten feet high, the chimpanzees calmly shuffled toward a termite mound for a snack. As the fire approached them, the chimpanzees didn't retreat—they stood and filed *toward* the flame front. Pruetz fled.

Pruetz had joined this group of chimpanzees because, unlike their kin in nearby forests, they live in savannah grasslands. These grasslands were similar to those that early humans would have adapted to when, about three million years ago, climatic changes shrank the forests they inhabited. Among these chimpanzees, Pruetz was looking for clues to our evolutionary past. And she found one when, after the fire, she reunited with her chimpanzee troop and discovered them unharmed, foraging through the charred aftermath.

Many animals that evolved alongside humans were attracted to wildfires. Some birds of prey have even been observed intentionally spreading fires by lifting flaming brands in their talons, dropping them in unburned fields, then eating the rodents that flee. Before the smoke settles, bird activity increases tenfold. On the ground, zebras and warthogs are often the first to appear, digging into the ashes to uncover exposed tubers. Soon after, elephants traverse the blackened earth, dispersing fruit seeds with their dung and creating corridors of orchards along their migratory pathways. Eventually, the burned area fills with herds of antelope and wildebeests. They graze new grass shoots, which are made more nutritious by the ash.

But Pruetz and her colleagues believe humans, our ancestors, and our evolutionary relatives stand alone in our ability to navigate active fire events. Not only can chimpanzees control their fear responses, calmly monitoring wildfires, but they can also predict how fires will spread. This is a complex process, requiring the chimpanzees to understand how flames respond to different plants, the steepness of slopes, and weather patterns. They do this so effectively, Pruetz told me, that she now tells her student researchers never to leave the troop in the event of a wildfire.

This cognitive capacity to navigate wildfires would likely have been present in the earliest hominins, providing advantages that were essential for the emergence of our species. Natural flames made predators visible, opened landscapes for walking, and even began cooking for us. Wildfires made seeds and nuts easier to digest, while softening meat and eliminating pathogens. New foods became available, bulbs and rhizomes that were inedible raw but, when roasted, provided abundant nutrients. Nature, however, is rarely a good chef, so hominins would have eventually learned to intervene, dragging carcasses from flames before the meat was charred or tossing roots into smoldering stump pits for tenderizing. In other words, long before anyone built organized hearths, our ancestors were likely already becoming a uniquely fire-adapted species by following wildfires.

By one million years ago, the physical effects of our coexistence with fire had transformed our bodies. Flames acted like an external stomach, digesting food before it entered our mouths. This allowed our teeth to shrink and guts to compress. Our bodies directed the energy saved from digestion toward our brains, which doubled in size, allowing us to develop more complex social structures, tools, and language. By half a million years ago, we had developed a genetic mutation, present in no other primate, that allowed us to safely metabolize carcinogenic toxins present in smoke and charred food. It's not so much that humans domesticated wildfires, but that wildfires domesticated us.

Eventually, though, humans didn't just follow wildfires. We nurtured them, kept them burning, and spread them wherever we traveled. We used fire to shape the land. The land adapted, growing plants that thrived with flames around our settlements. By sixty thousand years ago, when humans dispersed across the earth, our journeys could be traced by plumes of smoke.

Then, just a few centuries ago, a small group of humans began putting these fires out.

THE FIRES THAT HAD BEEN SUPPRESSED WHEN I FIRST ARRIVED IN California were different from those that occurred at any other time in human history. Driving up the coast from Los Angeles, I entered the burn scar of a megafire. For weeks, the road had been blocked by flames and debris, isolating Santa Barbara, our new home, in red sunlight and ash. This fire had taken the state by surprise, igniting in December and consuming fifty square miles in a matter of hours. At some points, the fire had spread at a rate of an acre per second, killing one firefighter, destroying over a thousand buildings, and forcing an exodus of one hundred thousand residents.

Cultural histories, economies, and ideologies converge to shape a region's ecological character, from which flames emerge. Driving toward Santa Barbara, I wondered what this perspective could reveal about California's fires. But I also wanted to get close to the flames, to understand how people navigate new scales of destruction—emotionally, physically, and tactically. Shortly after I began a PhD in the Anthropology Department of UC Santa Barbara, I became a wildland firefighter to see what I could learn on the front lines.

I also saw wildfire as a way to pay the rent. Within the first week of a multi-month hiatus between academic jobs, I became aware that I was an overeducated, unemployed millennial living in an overpriced garage. So I called the Forest Service hotshots. I had heard locals describe the hotshots as the Navy SEALs of wildland firefighters. The hotshots laughed at my gumption, but they gave me the phone number of a beginner crew.

I called the number. The man who answered was on his way to a wildfire ripping through the suburbs of Los Angeles. "This job is painful," the man warned. "You have to be able to push yourself harder than you ever thought possible."

I enjoyed pushing myself and traversing rough terrain. A few years before, I had hiked the length of the Hashemite Kingdom of Jordan. A few months before, I had crossed Bosnia from Montenegro on foot. I was pretty confident I could fight wildfires. The man hired me.

I was less confident that firefighters would take a book nerd seriously, so I trained hard, cycling through the Sonoran Desert, hiking along the Continental Divide, and running dozens of miles through Santa Barbara's coastal mountains. I trained so hard that when the crew's forty new hires raced up a mountain, I found myself in the lead. Hiking speed, I learned, is an informal metric that firefighters use to estimate how long they can work before collapsing.

Another would-be firefighter huffed to the summit behind me. "You gonna be a sawyer?" he asked, after catching his breath.

"A what?"

"A sawyer. The guy with the chainsaw. You hike fast. You should do it."

I had expected a hose, but when I learned that sawyers are paid an extra dollar per hour, I took the saw. My job was to lead the fire crew, cutting brush out of the path of the flames. The rest of the squad would follow with hoes and axes, chopping, scraping, and scratching until a solid line of mineral soil separated the approaching fire from the surrounding vegetation. Wildfires are slowed by airdrops of retardant and water, but they are contained and extinguished by lines of humans wielding tools.

While we trained, in 2020, unprecedented wildfires spread across the planet, but California experienced a tense calm, like a ship in the eye of a hurricane. Several members of the Los Padres Hotshots, Santa Barbara's local crew, were dispatched to Australia, where megafires killed thirty-four people, incinerated around a billion animals, and consumed almost one hundred thousand square miles. Then the entire hotshot crew was called to the Arctic, where megafires rolled through land that had perhaps never before been subject to these kinds of flames. Meanwhile, a drought in the Amazon sparked fires that erased thousands of miles of rainforest. As California simmered in the worst drought in over a thousand years, the planet's flames seemed to be circling, drawing nearer.

In August, megafires struck the coast.

BEFORE A HEAT WAVE TRANSFORMED IT INTO A MEGAFIRE, THE WILD-fire in Big Sur was standard, just one of more than sixty thousand that ignite in the United States each year. We fought it as we would any other. The first day, we approached the fire before dawn, marching along a dirt road into a redwood forest where condensed fog dripped like rain from the canopy high above. A mile in, we found the fire creeping close to the ground through the underbrush—a small corner of its slowly expanding thirty-thousand-acre perimeter. Our job was to solidify this perimeter, stopping the expansion by carving a line along the fire's edge.

I walked to the edge of the fire, put in my earplugs, pulled the chainsaw ignition cord, and began cutting through the wall of undergrowth. Colin, my saw partner, danced around me, catching the falling branches and bushes and hurling them away from flames that lapped against our boots. The rest of the crew followed, three saw teams widening my path and ten others with tools scraping the path to soil.

Through my earplugs, the scream of the chainsaw became a drone that blurred the hours until time became measured by tanks of gas. By the end of the first tank, when the chainsaw sputtered to a stop, sweat had soaked through the leather of my boots and Kevlar chaps. I called for fuel and dropped to my knees to open the chainsaw, chugging water as Colin pulled fuel canisters from his pack. Firefighters, like other endurance athletes, try to keep their packs light and tight. Bulky packs are torture when tunneling through thick brush—especially for sawyers, who constantly contort their bodies to reach the next branch, and their partners, who wrestle the vegetation. Every ounce of weight is extracted tenfold in sweat. Gear undergoes a meticulous selection until the packs are filled with only the essentials: gasoline for the chainsaws, six liters of water, enough food to metabolize the water, headlamps in case we worked all night, and emergency shelters the shape and weight of bricks. Even after this winnowing, packs often weighed fifty pounds. It took Colin about a minute to refill my chainsaw, which weighed twenty-five pounds with fuel. I hauled my pack onto my shoulders and kept cutting.

By the second tank, we had followed the edge of the fire out of the forest and into overhead brush that bristled from an almost vertical slope. Colin pressed his hands to my buttocks to steady me as I lifted the chainsaw overhead, cut the base of a bush, gripped the stump like a ladder, and pulled myself up. Handling a chainsaw in this terrain felt like trying to mountaineer while cradling a feral animal that would bite off my hand if I made a wrong move. When the fuel ran out again, the silence startled me. I realized I was gasping for air.

By noon, when the sun burned through the fog, the rhythm of the fuel tank changes began to blur. The flames lengthened and the smoke thickened. My forearms cramped, contorting my hands into claws that I wrapped around the saw trigger to keep cutting. I realized I was running out of nutrients, so Colin began refueling me as well, popping salt pills into my mouth before filling the gas tanks.

When the sun began to set, the temperature dropped, and the fire settled back into the land. We hiked down our fireline to our trucks. Hands trembling on steering wheels, we caravanned south along Highway 1. The sun melted into the water, pooling along its surface. The horizon fused in smoke until everything was a wall of purple and gold. I noticed whales breaching within those colors, blowing droplets into the dusk.

I was half asleep by the time we arrived at the incident command center, a smattering of military tents spread along the only stretch of flat ground. The tents brimmed with commanders perusing maps, nurses bandaging blisters, and cooks ladling slop for the five hundred or so firefighters who needed to be fed. Colin entered the camp to fetch our food, which I ingested mechanically, too tired to taste anything. After, we drove down the road to sleep on the ground.

The heat wave was announced on a handheld radio in the predawn darkness of the eleventh day, when the fire was almost contained. My fingernails were falling off from the constant vibration of the chainsaw. I was sharpening the chainsaw on the tailgate of a truck on the shoulder of the empty highway. Colin, my saw partner, had just appeared through the fog, bearing two cups of coffee, and now stood beside me to prepare our daily mix of vitamins. The crash of waves almost swallowed the staticky radio voice announcing that temperatures would rise to 120 degrees Fahrenheit, the hottest ever recorded in the region. We were to expect unprecedented fire behavior.

We paused and looked at each other. He shook his head, grimacing like he was about to be punched. If climate change were tracked by firefighters' weather briefings, that voice may have flagged a tipping point. To Colin, it sounded like pain. To me, it sounded like a death sentence.

THE DAY OF THE HEAT WAVE, I WAS BACK ON THE RIDGE WITH FINNEGAN, my rival sawyer. We were watching the fire. I noticed a figure moving through the smoke. The figure shimmered like a mirage, marching up the fireline, silhouetted by ocean and flames. Finnegan straightened beside me and wiped his face, smearing dirt through his sweat. The rest of the crew rustled as well, struggling to look presentable within the scant shade of the bushes. Squinting, I realized the cause of the commotion: the approaching figure was a hotshot.

Hotshots are firefighter legends, the Special Forces of wildland firefighters. On the beginner crew, we rarely saw them because they fought the most violent sections of the fire and slept near the flames, as opposed to the main camp where the rest of us slept. Occasionally, we heard their chainsaws screaming through the smoke. When we caught sight of them, they were easy to identify, moving with a clipped stride, downturned faces, and hunched shoulders. Their strange gait told a story—they were more comfortable on slopes than flat ground, they weren't seeking attention, and they couldn't stand straight after months bearing fire packs.

By some measures, being a hotshot is among the most difficult jobs on earth. Hotshots don't just mountaineer; they mountaineer with chainsaws, and they do so in thick smoke, often in unprecedented heat, along the edges of the most extreme conflagrations in recorded history. In the mountains of California's Sierra Nevada range, hotshots sometimes train alongside Olympic athletes because their jobs require an equivalent level of tactical athleticism. They need to be able to hike for hours, cut line for miles, command helicopters and airplanes from their radios on the ground, and maintain a constant awareness of all the shifting conditions that could change the fire's spread. Their lives depend on these abilities.

I wanted to be a hotshot. Finnegan wanted to be a hotshot too. But there were only one hundred hotshot crews in the United States, twenty active in California, and each crew had only a few openings each year. For the sake of comparison, roughly two thousand men and women serve on hotshot crews, about the same number of people in the NFL. The competition is stiff.

The hotshot was here because trouble was brewing. A mile above us on the fireline, a falling boulder had injured a man's leg. A hotshot crew had carried him to the summit for a helicopter evacuation. From the ridge below, we watched the helicopter circle the coniferous crown of the mountain, searching for a landing pad before eventually buzzing away. A voice on the radio announced that the pilot had abandoned the mission because the landing was too dangerous. The sun was sinking, the fire was growing, and the injured man was trapped in the fire's path. Someone needed to carry evacuation equipment to the summit so the team could carry him down. When the hotshot stopped to talk to our crew leader, we knew it would be one of us.

Our crew leader pointed in our direction. Finnegan's chest swelled and he stepped forward, but the crew leader shook his head, gesturing at me. The hotshot waved me over. His shirt, once the standard mustard-yellow of the Forest Service, was now black with soot and crystallized with sweat. His face matched his shirt, with a mustache so dirty it looked pasted like putty. He was breathing heavily from the hour-long hike from the coastal highway, sweat streaming from his chin, but his eyes darted with humor as he appraised me. At a certain level of physical suffering, the pain becomes almost comedic. Standing in the heat, the hotshot seemed ready to laugh.

"Heard you're a good hiker." He grinned, or grimaced, or something in between. "Think you can make it to the top of that mountain?"

I nodded because I didn't trust myself to speak. This hotshot had been fighting wildfires for four months straight, sleeping in the dirt, probably with a single pair of underwear, while burning the caloric equivalent of a marathon every day. This was my first fire, and I had been fighting it for less than two weeks, but my body was wrecked. My feet were torn and oozing within my elk leather boots, and every inch of my skin was a rash of poison oak. Hours before, I had been incapacitated by muscle cramps. But I wanted to be a hotshot. This could be my only chance to prove myself.

Besides, a man was injured on the mountaintop, and he needed help.

I took the medical pack, and I began to run.

THE FIRST FEW STEPS WERE THE MOST DIFFICULT. WITH THE EYES OF the hotshot on me, I hobbled awkwardly to prevent my legs from cramping until my muscles loosened and I could lengthen my strides. The equipment was lighter than my fire pack, but it stung my shoulders where they chafed. My feet were flayed from the leather. The only antidote to the discomfort was to return to the level of exhaustion where the body becomes numb.

I gained speed down a short decline, digging my heels into the dirt and looking toward a slope that formed a wall ahead. Brush piled on my right, the ocean expanded to my left, and flames licked up the slopes. *Breath, step, heartbeat.* I slipped into a pattern of movement. The temperature in Big Sur had reached 119 degrees that day, and for every degree the body's internal temperature rises, the heart quickens. At 145 heartbeats per minute, complex motor skills begin to diminish. At 170 beats, you enter tunnel vision, start to lose depth perception, and your hearing is restricted. In extreme heat, core body temperatures can simmer around 106 degrees. At around 180 beats per minute, rational thought flees, and you enter a netherworld of instinct. "That is where we live," a retired hotshot once told me, "on the blurry line between consciousness and unconsciousness, where one more ounce of energy will push us over a dangerous threshold. It is our home. And we should thank this mountain, because it takes us home."

I hit the uphill slope and the run became a climb. With one hand, I swung the hoe overhead, buried it in the dirt, grabbed a rock with my other hand, and pulled myself up the fireline. I climbed past the Scorpions, a crew that splits time between farmwork and firefighting. They blared mariachi music from their phones and cheered me on. I passed another crew, whose members clapped me on the back and told me I was close. My palm was bleeding, gashed by a rock, but I barely noticed. My vision tunneled toward the pine trees above, which seemed to grow with every step.

Nearing the summit, the fire column glowed in the sunset like it burned from within. The mountains were black pyramids amid swirling colors. My thoughts felt detached from my body. They were teetering on delusion. When the helicopter approached again, I wondered if it might hoist me up and carry me away. Instead, it made one more futile pass. I was close enough now that I could see the trees quiver before the helicopter sped off.

The hotshots waited for me on the mountaintop. The injured man lay on the ground with a pale face, a mess of dark hair, and frightened brown eyes. With a nod, a hotshot pulled the medical equipment from my back. Another radioed to their leader that I had arrived. My own crew was a collection of specks below. I sat on a log, trying to steady my breath, watching the horizon expand and contract with my pulse until everything slowed.

But there was no time to waste. When the hotshots hefted the injured man onto a stretcher, I plunged back down the slope, into darkness, into a hallucinatory stupor where I joined my crew to cut an evacuation route for the hotshots to carry the man down through the forest. In the night, my headlamp cast shadows into monstrous shapes that prowled the trees. I tore through the branches and they tore back at me, until I landed in an orb of white light where the ground was flat and open and hard underfoot.

The leader of the hotshots stood in the road. He was silhouetted by the glow of headlights. His face was soot. His mouth was set in that same amused grimace. His eyes were focused on me. When I approached, he asked if I had ever considered becoming a hotshot.

I tried standing straight, but my body wanted to curl in on itself, pulling me toward a standing fetal position. I couldn't imagine doing this nonstop for six months. It took the last of my strength to tell him yes, I wanted to be a hotshot.

The hotshot shook my hand and invited me to interview.

CHAPTER 2

IN JANUARY, FOUR MONTHS AFTER FIGHTING THE MEGAFIRE IN BIG Sur, I drove toward the headquarters of the Los Padres Hotshots to interview for a position on the crew. My old Prius clanked over the shrouded coastal pass before heading down into a dry fold in the mountains. These mountains, the Santa Ynez range, are among the most recent geological formations in the Americas, driven from the sea just four million years ago by a collision of tectonic plates. The slopes of this range are too young to have been smoothed by erosion, so they frequently blur distinctions between mountains and cliffs. Heat rises, and so do flames, spreading most rapidly upslope. In these dark canyons and fanged peaks, fire accelerates exponentially, catching and climbing and jetting over ridgelines. Embers from the eruptions rain into adjacent valleys, barreling forward in search of oxygen.

The youth of this range alone would be enough to create formidable fire behavior, but a chance kink in the fault line tipped the scales toward extreme. When the tectonic plates collided, the land buckled from east to west, making these mountains one of just two transverse ranges in North America. Every year, late in autumn, at the height of Southern California's dry season, air flows southwest across the Great Basin, becoming hotter and drier as it rushes over the Mojave Desert. In a normal mountain range, the air would crash against the slopes and disperse. In Southern California, the east-to-west valleys suck in the air, heating it as it howls toward the coast, often at over one hundred miles per hour. When the brush senses the impending blast of wind, the plants transfer moisture to their roots, leaving branches bare and brittle, as if begging for ignition. Their leaves become glossy with flammable waxes and resins, poised to transform any spark into a landscape-scale explosion. In Southern California, the mountains consume fire as much as fire consumes the mountains.

The region's Indigenous Chumash inhabitants have historically said a fire demon named Asiqutc inhabits these peaks, an old witch who breathes sparks into the wind she casts with a woven fan. The Spanish called this demon La Quemadora. Scientists labeled her a foehn wind. In Santa Barbara, the demon became the Sundowners, because the winds often pour over the front range at night, breathing flames toward the city lights on the coast. When the fire demon fans the flames, no human force can stand in her way.

But this was precisely the job of the hotshots.

※

"YOU DON'T NEED TO KNOCK!" SOMEONE SHOUTED FROM INSIDE THE hotshot office. I paused before my knuckles could hit the door again and stepped inside. The hallway was cluttered with portraits of the Los Padres hotshot crew. Decades stretched along the walls. One man seemed to be in each photo, a tall figure with long black hair that became streaked with white as I moved through the hall.

There, behind a wooden desk, was the man himself: Aoki, the

leader of the Los Padres Hotshots. He was sending emails with untied hair falling down the back of his chair. Edgar, his forty-year-old captain and second-in-command, sat behind another desk, under a plaque emblazoned with a prayer for dead firefighters. Their desks were in a power formation, looming side by side in front of a rickety chair that looked like it was meant for me. I took a deep breath and sat. Aoki was widely regarded as the most experienced hotshot in California, a state that claims the most skilled firefighters in the world.

Aoki ignored me. Edgar appraised me with a reptilian stare. He looked like a military guy—crew cut, shiny face, and muscles bulging from his T-shirt. He didn't blink until he looked down at a paper and started ticking through questions: my hobbies (hiking, camping, and lots of exercise, which my new fire-engine friends had coached me to say), athletic history (football and wrestling, albeit a decade ago), and fire experience (two years on a beginner crew, which didn't impress Edgar a bit).

"Anything else?" Edgar asked, laying down the paper. "Do you have any mechanical skills? Carpentry? Welding? Anything else you'd contribute to the crew?"

I panicked. "Um, I garden."

Aoki, sipping coffee and typing, looked up and chuckled, spectacles balanced on the tip of his nose. Edgar narrowed his eyes. Without a word, he made a note, clacked his sheaf of papers on his desk, and filed them away.

Aoki stood and stretched, body unfolding like a predatory bird. He finally fixed me in his gaze. "All right, Thomas," he said. "I was thinking about taking a little walk up the hill. Want to come?"

Under normal circumstances, the invitation may have seemed a polite, if quirky, facet of forestry culture, an opportunity to escape the office for the jaunty conversational space of the woods. But I knew better, thanks to my gossipy fire-engine friends from the beginner crew. Leaning in, eyes wide, they had warned me that *this* was when the interview would really start.

None of the fire-engine guys had been a hotshot, but they had gathered enough rumors and stories to provide me with a mental map. I needed to submit the formal application through the online government portal, but that didn't really matter. What mattered was that I call the hotshot station in January, after they would have recovered from fire season. I should dial at precisely 7:50 a.m., after the hotshots would have arrived but before they started their day. The engine guys laughed at my plan to request an in-person interview. "Play it cool," they said. "Like, 'Hey, I was thinking about cruisin' by and checkin' out the station. You around?'"

I was also to moderate my biological processes for several weeks before the interview. Be in shape and hydrate obsessively, obviously. Wake early, to be alert. Eat pasta the night before, for the carbs. And make sure to use the restroom before I leave home, because I would look weak if I soiled myself during the interview. It had happened. Prompted by my puzzled expression, they explained that the office session was just a formality. The real interview was a hike.

As my fire-engine friends advised, I had worn my elkskin boots and, before entering Aoki's office, had positioned a fifty-pound rucksack outside the door. I was to accept Aoki's invitation, don my pack, then hike up a nearby mountain until I was on the verge of losing consciousness. Even pushing fifty, whispered the engine guys, Aoki could hike people to death. Barefoot. His lungs held double the capacity of regular humans. He survived on coffee and chicharrones. If I wanted to get on the crew, I should push myself to the brink of blacking out but, throughout the hike, remain cognizant enough to answer his questions. If I made it to the top, I had to meet his gaze. "He'll look into your eyes to see if you seem like you could slam line for the next twelve hours," an engine guy had warned. "That's what hotshots do."

I felt I had botched the office interview, but I was prepared for the hike. Aoki grunted with approval as he stepped past my pack into the morning sun. He didn't wait for me to prepare. The weights in my backpack clanked as I scurried to catch up, trailing him around the barracks and along the edge of a sand volleyball pit.

To make conversation before the climb, and to show Aoki I had done my homework, I asked about the hotshot volleyball rivalry. The Los Padres and Tahoe Hotshots contest each other for the status of the nation's gnarliest crew, and volleyball was a space to battle out these competing claims. "We always used to kick their asses," Aoki replied. But nowadays there wasn't time to play. All the hotshots in California fought fires from May until November.

Past a supply shed with a tin roof and chicken-wire walls, between split logs and a chopping ax, we stepped onto a path that climbed into a deciduous forest. "It's just a walk," Aoki reminded me, hearing my labored breath, "and we haven't even started yet." We crested the knoll into an open park of mobile homes, where Edgar's girlfriend waved from a porch, before crossing a road to a rusted sign: Snyder Trail.

Each Forest Service fire station in California is associated with a trail, and each trail has a reputation. The one where I had trained with my beginner crew was notorious: Laguna, an eight-tenths-of-a-mile sprint up a thousand feet of elevation. Wildland firefighters gauge each other's prowess by prying into hiking times. A good time up Laguna was 24 minutes. I had seen a firefighter finish in 21 minutes before he collapsed and was airlifted to a hospital. My best time was 19:43. When Aoki was a beginner, legend claimed that he had climbed Laguna in 13 minutes.

Snyder Trail was a different beast. While Aoki stopped at the trailhead to fiddle with his watch so he could time me, I watched the clouds of Santa Barbara spilling over from the other side of the ridge two thousand feet above. Their tendrils grasped skyward before evaporating in the sun. The front range cradles the wet ocean air like hands cupping water. On the inland side, where we were, it was hot and dry. Snyder Trail climbed for almost two miles. Most hotshots could finish in thirty-five minutes. "Let's go," Aoki said when his watch beeped. "Remember, it's just a walk."

I didn't begin with a walk, but with a lumbering jog under the sycamore canopy. My heart was already racing, more from nerves than exertion. Aoki stayed on my heels. He waited until the trail steepened out of the trees before he spoke.

"So, why do you want to be a hotshot?" he asked. His voice was level, like we were sitting in a coffee shop.

"I want to learn about fire," I replied, between breaths. "And everyone says you guys are the best." I knew why he was asking. Many people use being a hotshot as a stepping-stone to a municipal crew, where people earn triple the salary working half the hours. Even worse would be an aspiration toward Cal Fire, the state's own fire agency, where firefighters grow plump on full meals and hotel beds. The very bane of hotshots' existence as a federal resource had become a source of scrappy pride: low pay, bad food, sleep in the dirt. Aoki was silent for a moment as we crested the hill, which allowed me to recover before we hit a series of switchbacks into another grove of trees.

"Hotshots don't take vacations," he explained in the same parlor tone. "Expect no holidays, no birthdays, no weddings, no anniversaries for six months." Aoki had recently missed his son's high school graduation. "We've all missed those. The offseason is yours. The fire season is for the crew."

I told him I'd already cleared my calendar.

"Got an old lady at home?"

"Yep."

"What would she think if you got the job?"

"She'd . . . probably . . . be more excited than me." It was becoming hard to squeeze out words between breaths, but Aoki stayed silent, so I kept talking. "The way we see it"—two breaths—"I'll be home four days a month"—two breaths—"so we'll go on four dates a month . . . which is more than we do now."

That earned me a laugh. Soon we passed some invisible waymark, and he said we were halfway there. I was glad he was behind me. I'm a mouth breather, so drool was hanging from my chin. We left the trees and emerged onto an exposed slope that seemed to amplify the heat. The trail became hard and cracked underfoot.

"You know," Aoki continued, "this job has ended a lot of relationships."

I grunted to show I was listening. I was getting tired, focusing on opening my hips to maintain my pace, but the cognizant part of my mind wondered what he was getting at. It would be nice if Aoki cared about the health of my relationship, but it seemed unlikely.

"The work is demanding," he said. "The physicality is only the start of it. It's psychologically taxing. And it's a hard thing to be going through the emotional turmoil of a breakup on top of the physical intensity. It puts everyone at risk."

I understood. For Aoki, the health of our relationships was about the health of the crew.

Small groups aren't just collections of autonomous individuals. They become an organism. Like a virus, low morale can begin with one individual and spread, sapping the physical capacity and mental fortitude of the entire group. A single person lost in the brambles of their mind can snare everyone in the thicket. On the fireline, as in war, a loss of focus is measured in lives.

I told him I was almost thirty and had been with my girlfriend for seven years, so we knew our capacities. I also told him we had recently adopted a dog, which was better company than me. Aoki chuckled again, I assumed because of the dog. More likely, he was amused that I thought I knew what I was getting into, but he let the matter drop.

I felt myself nearing that physical terrain where slobbering gives way to tunnel vision. I had a choice: get to the top faster or maintain conversation. This was, I realized, a quick simulation of life on the fireline, a test of embodied attunement. Hotshots are specialists at keeping their bodies on the brink of a precipice where physical exertion teeters on mindless instinct. They learn to inhabit this edge while keeping their minds sharp, focused, and cognizant of their surroundings. When the workplace is filled with crashing trees, rolling boulders, and bursts of flame, the ability to skirt this precipice without slipping becomes a high-stakes art.

I allowed myself to slow as the landscape unfolded around us, a jagged panorama of blue slopes and red cliffs and golden summits. The beating sun made me notice that brush was conspicuously absent from this trail. Remembering the blackened hills I had crossed when approaching Santa Barbara for the first time, I asked Aoki if that megafire had reached these slopes.

"Not that one," he replied, but he pointed to an area that had burned in 1995, another in 2006, and several more the following decade. Each burn scar, he said, was associated with different patterns of vegetation and coloration that he could interpret for flammability. If a new fire were to ignite, Aoki would read these burn scars to predict where the new ignition would spread. But for me, squinting through sweat, all I could see was a rolling spectrum of beige.

I asked how he predicts how wildfires will behave.

"Really, all you need to know is S-290," he said. This answer jarred me. Perhaps I had been expecting something more mystical, a sixth pyro-sense branded into hotshot bones; S-290 is just a standardized Intermediate Wildfire Behavior course. "That's the foundation," he continued. "You need to know how fuel, weather, and topography interact with fire."

Aoki asked if I had completed the course. I had taken the online version. I remembered a few details from my computer screen: the robotic voice, simulations of swirling wind, and mountains animated with flickering fire. I also knew that California's conflagrations were beginning to transform fundamental assumptions of how fire works. Scientists around the world were racing to understand flames that cast ash forty-two thousand feet into the sky and generate winds that rip trees from their roots. Aoki's response was sanguine, yet sincere. Searching for burn scars across the mountains, I was looking for something I didn't know how to see, that I wasn't certain was even visible. Hotshots discipline the very rhythms of their thoughts and senses into fire suppression tools, until a mountain becomes a burn scar, a river a fuel break, a cloud an omen, a breeze a hazard. For hotshots, fire suppression isn't just a way of life—it is a way of being.

I wasn't going to get my answers on this hike. This sort of knowledge could only be gained on the fireline. I stopped talking and hiked faster.

ABOUT A CENTURY BEFORE MY INTERVIEW WITH AOKI, ANOTHER ANthropologist embarked on a journey to understand fire in California. But it wasn't conflagrations that puzzled Omer Stewart when, during the 1930s, he pursued his PhD at Berkeley. It was their absence. There were no conflagrations; or, at least, none near the scale of today. The absence of flames in California, for Omer, contradicted everything his research had led him to expect—that California should be burning just about everywhere, every year.

When Omer approached his mentor, Alfred Kroeber, with this mystery, he was in his late twenties, clean cut with squinted eyes and a disarming smile. Kroeber was at the time considered the world's foremost academic expert on California's Indigenous cultures. Omer convinced Kroeber to allow him to read his personal field notes from his research with the Yurok tribes in the north, toward the border of Oregon. Within those pages, in scribbled handwriting smeared by rain, Omer found repeated references to the ways Yurok people had managed their land just decades before. They burned in the meadows, in the forests, in the oak stands and redwood groves. "They burn," Kroeber had written, "far up in the mountains."

Intrigued, Omer traveled one hundred miles north of Berkeley to Clear Lake, home of several Pomo tribes. The traditional lands of the Pomo people include much of today's Mendocino, Lake, and Sonoma counties, a total area larger than Connecticut. Interviewing Pomo people about how they managed their land in generations past, Omer was told that they "had intentionally set fire to *all* the fields and forests."

This puzzled Omer, because it contradicted widespread assumptions that Indigenous people had done little to change landscapes before the arrival of Europeans and Americans. "It is often assumed," writes historian Stephen Pyne, "that the American Indian was incapable of greatly modifying his environment and that he would not have been much interested in doing so if he did have the capabilities." Omer suspected something to the contrary—that Indigenous people had not just altered ecosystems, but were widely responsible for creating the natural abundance that colonizers encountered, interpreted as wilderness, and exploited. Fire, Omer thought, was the missing evidence.

At the time, this was a revolutionary idea. It challenged the very concept of untouched wilderness, which lay at the core of the colonial encounter with the New World, and, with it, the claims white settlers had made to that land. In 1786, a French seafarer recalled being awestruck by California's "inexpressible fertility." The southern coast was a park of ancient oak trees towering above a carpet of edible chia grasses. The central mountains were covered in brush that provided fruits and berries for grizzly bears, black bears, and deer. The rolling alpine forests of the Sierra Nevada range were open and dappled with sunlight that nourished the forest floor with herbs and flowers. Shoals of fish darkened the ocean, herds of elk shrouded the plains, and birds flocked in the sky by the millions. In 1849, a young American traveler wrote to his fiancée, comparing this ensemble of life to "some new-created world."

If this were a new-created world, then Indigenous people had no place in it. At best, settlers and scientists alike considered them with condescension, as ecologically impotent, and thus icons of conservation to be held against the excesses of American capitalism. "An Indian took pride not in making a mark on the land," wrote historian J. Donald Hughes, "but in leaving as few marks as possible: in walking through the forest without breaking branches, in building a fire that made as little smoke as possible, in killing one deer without disturbing the others." The possibility of beneficial influences, such as enhancing the numbers and diversity of other species, was seldom considered.

More often, the impacts of Indigenous people on the land were assumed to be negative. John Muir, the founder of the American conservation movement, first entered Yosemite Valley in 1868, seventeen years after state militias violently removed the Southern Miwok people who lived there. He described the land as "abundant, divine, gushing, living plant gold, forming the most glowing landscape the eye of man can behold." He also viewed the valley's Indigenous people as intruders on this state of nature. Muir supported their removal and advocated for the suppression of their flames.

Omer suspected that both narratives—the ecological eunuch and the intruder—were farcical at best. At worst, the narratives served as ideological pillars that propped up white entitlement to land theft. Moreover, scientists had enabled this farce by neglecting the role of intentional, human-caused fires in shaping landscapes.

So, for the next decade, Omer endeavored to set the record straight. He traveled the country, descended into archives, lurked in libraries, and harangued historians. He gathered diaries of Spanish explorers, letters from homesteaders, and reports from state and federal officials. The documents he compiled spanned three centuries. They cluttered his desk, spilled from his filing cabinets, piled up in office corners.

Over the years, as Omer organized his findings, he found that the story of fire, Indigenous people, and the land was not restricted to California, but spanned most of the United States. Settlers and colonizers recorded that the entire Eastern Seaboard from Maine to Florida was annually filled with smoke and flames. In 1609, in the northeast, an Englishman noted, "They meett sum 2 or 300 togither and euery one with a fier stick." In Virginia, forty years later, English travelers described a forest where, "as almost everywhere, the Indians followed the custom of burning in the fall." The woods were so clear that they could see to the horizon, with trees spaced widely enough for carriages to ride through. Wherever sunlight burst through the canopy, strawberries and grapes grew in abundance.

Omer followed this paper trail inland to Kentucky, Louisiana, Mississippi, Texas, Arkansas, Missouri, Iowa, Illinois, and Minnesota. In each region, there was evidence that everything from the hardwood forests of the north to the shortleaf pines of the south and the prairies of the Great Plains had been shaped by fires. In Michigan, a settler in 1835 described flames curling along the ground, licking through the grass, and tumbling "like the breakers of the sea." In Wisconsin, a man wrote that "these fires were not conflagrations of catastrophic proportions which destroyed the primeval forest . . . but rather periodic and ecologically normal events."

Many were enthralled by the flames Indigenous people set to the land. In 1842, George Catlin, an artist, watched tribes burn the prairies of western Missouri, near Kansas City. "These scenes at night become indescribably beautiful," he wrote, "when their flames are seen at many miles distance, creeping over the sides and tops of the bluffs, appearing to be sparkling and brilliant chains of liquid fire . . . hanging suspended in graceful festoons from the sky." Indigenous fires shaped the prairies that stretch all the way from Canada to Texas.

In the Rocky Mountains, the story was the same. The Salish tribes have a Sx'paám, a "Fire Setter," who, historically and today, has been revered for their knowledge of fire's effects on animals, plants, and landscapes. Another interpretation of Sx'paám is someone who kindles fire, here and there, over and over, indicating a consistent practice of tending to habitats and species adapted to fire.

Across the Great Basin, into California, burning was even more common. In 1769, a member of the first Spanish overland expedition from San Diego to San Francisco reported that he had difficulty finding unburned grass for his horses to graze. The rest was black, lit by coastal Chumash tribes.

North of San Francisco, in the great coastal forests, explorers noticed that various tribes used fire to prevent trees from encroaching on meadows. This confused the explorers until one party got lost in the woods. For ten days, they walked without "the sight of any living thing that could be made available or useful for food." They were starving. Two pack mules died of hunger. Then, ascending a rocky embankment, they reached a fire-maintained prairie and found "on one side . . . little knots of deer, on another, a large herd of elk." Before reaching any of this game they discovered and shot five grizzly bears and feasted for the next several days. Indigenous people used fire to create a mosaic of ecosystems that ensured a steady supply of food.

For Omer, that was intuitive. What wasn't intuitive for most scientists at the time was the idea that Indigenous people didn't just mimic natural fire patterns, but created their own patterns of cultural burning, and that these cultural burns had become the defining feature of California's landscapes. When humans first entered the coastal area of California some thirteen thousand years ago, they found an environment that was already combustible. Over the course of the next millennia, instead of fighting fire, California's Indigenous people gradually maneuvered this flammability to their advantage. By the time Europeans arrived, fire was the crux of California's ecosystems, and these ecosystems were central to Indigenous economies. Without the right kinds of fire, applied with knowledge and skill, everything would collapse.

The forms of fire that emerged as a result of Indigenous management are as diverse as the people and ecology of Indigenous California. The Chumash people of the Central Coast burned patches of brush every fifty years or so, but also used frequent, light fires to create edible landscapes on the coastal plains. Ethnobotanists have documented that these practices not only limited the severity of wildfires, but also increased the abundance of foods, medicines, and building materials. Using fire, the Chumash cultivated a mosaic of hardwood forests, shrubs, and grasslands between the coast and the mountains.

Amid the Sierra Nevada mountains, Indigenous people were attuned to different ecological relationships. Many of the edible berries, roots, and greens of the coniferous forests only grew in spotted sunlight, so tribes burned the understory every two or three years. Their flames cultivated habitats for deer, opened migratory routes for elk, and provided pathways for bears, mountain lions, and wolves. They also burned to produce edible fungi: the coccora, sweet tooth, black morel, woodland cup, and others that grow best when stimulated by heat. These tribes, said Karuk ecologist Frank Kanawha Lake, "literally used fire from the coast to the highest alpine meadows."

According to Jan Timbrook, a California anthropologist, the region's Indigenous people used fire to produce food as effectively as Europeans used plows, supporting a population that ranged from three hundred thousand to half a million. They typically avoided sedentary agriculture not because they were unaware of the practice—they surely knew of it, as their trade routes stretched from the cities of the New Mexico Tiwa to the pyramids of Tenochtitlán—but because agriculture would have been less efficient. By tending the land with flames, they were able to support some of the highest populations in the Americas.

The use of fire, however, wasn't limited to food production. Tribes also burned to shape the broader ecological processes of landscapes. The Mono people of Northern California sprinkled fire in river valleys to reduce plant transpiration and raise water levels. Other tribes burned mountaintops to clear high elevation foliage. This tightened the snowpack so it would melt more slowly in the spring, reducing the risk of flooding while ensuring that water flowed throughout the summer.

Just as Indigenous burning connected fire to water and soil, it melded the biosphere with the atmosphere. For every acre that burns, smoke carries some forty trillion live microbes into the sky, dispersing fungal spores and bacterial cells across the landscape like ocean currents across the seas. These microbes settle, sink, and establish, hybridizing with local biomes. The smoke historically dispersed sunlight and cooled streams with enough regularity to assist the migration of salmon.

For Indigenous people throughout the Americas, burning was not just a matter of technical skill, but a way to tend their relationships with the land. "To us," writes Tarahumara scholar Enrique Salmon, "the land exists in the same manner as do our families, the river, and the sky." In many Indigenous societies before the European conquest, entire economies were maintained through reciprocity—a system of collaboration and exchange. A gift of food, basketry, or assistance was not merely an altruistic act, but a way to stitch together relationships, creating attachments and responsibilities among people, communities, and environments. Through reciprocal exchange, Salmon writes, "everything is woven into a managed, interconnected tapestry."

When the land includes everything, says Potawatomi botanist Robin Wall Kimmerer, society includes the environment, and the land becomes an active member of the social exchange. To maintain these connections, and to ensure that the land continues to share its productive wealth, people have been obligated not just to extract and harvest resources, but to provide something to the land in return.

For Indigenous people, fire has long been that gift. "We're told that's why the Creator gave people the fire stick," writes Kimmerer, "to bring good things to the land." She describes fire as a luminous instrument, a paintbrush for the landscape. "Touch it here in a small dab and you've made a green meadow for elk. Draw the fire brush along the creek and the next spring it's a thick stand of yellow willows. A wash over a grassy meadow turns it blue with camas. To make blueberries, let the paint dry for a few years and repeat."

Little by little, year by year, with each stroke of fire, California's Indigenous people transformed the region into something new. In the centuries before the arrival of Europeans, Indigenous flames and lightning fires together burned approximately ten million acres per year—more than double the acreage consumed in the worst fire season in California's recorded state history.

But the fires then were different. Under the Indigenous torch, these fires were rarely catastrophic. On the contrary, the flames of Indigenous California enhanced the biological diversity of the region. "What are now called 'fire-dependent' landscapes," writes anthropologist Adriana Petryna, "are, in fact, the product of over ten thousand years of fire use across ancestral territories."

In 1542, the glow of the landscape was noticed by conquistador Juan Rodriguez Cabrillo, standing at the helm of the first Spanish galleon to sail up California's coast. Squinting toward the horizon through churning seas, he saw a throb of rosy light. "*Una tierra del fuego,*" he allegedly told his crew. They were nearing a land of fire.

But this glow was not from flames. The light emanated from flowers that grew from the ashes, which swaddled the hills in such thick foliage that the petals became radiant, a beacon attracting the Spanish from twenty-five miles offshore.

❧

THERE IT WAS. OMER STEWART HAD PROOF THAT HIS THEORY WAS correct. Not only had Indigenous people, through fire, possessed the capacity to shape entire landscapes, but they had done so with a sophisticated understanding of ecological processes—an understanding that surpassed even Western science. He was poised to shatter American fantasies of untouched wilderness, the Indians of Eden, and the idea that fire exists as a destructive monolith. He began compiling his findings into a book.

Omer was in his early forties when he completed his book in 1952. He received an invitation from the New York publisher Alfred Knopf, who said he personally wished to consider the manuscript. Omer sent a draft, apologizing for the rough form and requesting suggestions. In less than a week, he received a scathing rejection. "Your manuscript as it stands is in perfectly impossible shape," wrote Knopf. "I am afraid that we can't see in it even the making of a book."

Six months later, Omer exchanged letters with a small anthropological press. Two years passed. They weren't interested.

A decade later, Omer sent the manuscript to a Florida publisher. The publisher lost the manuscript. After twenty-five years, he received a packet in the mail. They had found the manuscript in a filing cabinet but had lost any desire to publish it.

Omer was in his eighties when he submitted it a third and final time, six months before his death on New Year's Eve 1991. Over half a century had passed since he began his research; over forty years since he had first set pen to page. Again, he was rejected. The book was too outdated.

The world had moved on, and little had changed. Fire science had gained traction, with new technologies that could model fire behavior, predict its spread, and monitor its effects. But this science was all directed toward suppression. The science sought to understand the enemy so it could be vanquished.

Moreover, in those same years, resistance to Indigenous burning had only hardened. "It would be difficult to find a reason why the Indians should care one way or another if the forest burned," wrote an official of the California Division of Forestry in 1959. "It is quite something else again to contend that the Indians used fire systematically to 'improve' the forest. Improve for what purpose?"

They couldn't understand, or they didn't want to. Little had changed from Omer's childhood in the 1910s, when the district ranger of Klamath National Forest wrote to his supervisor that the solution to Indigenous burning was "to kill them off, every time you catch one sneaking around in the brush like a coyote." Another option, the ranger suggested, was to hire a missionary to convince Indigenous people to adopt Western theories about fire.

A Yurok tribal member who lived in the Klamath region, and who had collaborated with Omer's mentor, penned a letter of his own in response to the growing tide of suppression.

"You're going to regret this," the letter warned.

The words were prescient. No one listened.

CHAPTER 3

AFTER THIRTY-TWO MINUTES AND TWENTY SECONDS OF HIKING with Aoki, I reached the top of Snyder Trail. Edgar called me the next week to offer me a position on the crew.

Four months later, on the tenth of May, I stood on the lawn of the Los Padres Hotshots compound among the men I would spend nearly every moment with for the next six months. We formed a circle for introductions, the sun already blazing through the early morning mist.

When I introduced myself as a sawyer, the hotshots laughed.

I couldn't tell who laughed first because everyone looked the same, with matching forest-green pants and T-shirts, leather logging boots, and hats pulled low over sunglasses. I knew Aoki, with his long hair, and Edgar, with his glare. Most faces were clean-shaven, like theirs. The few that weren't shaven were extravagantly hairy, noses popping out from bushy beards. Several men were covered in military tattoos. One, a former Marine, had a pending felony charge for assault with a deadly weapon, his fists. Most were from California. One grew up in Nicaragua, another in Mexico. A few were surfers with salty drawls and triangular bodies.

The only characteristics that seemed to unite these men were a very high tolerance for pain and the joke that the new hotshot thought he was a sawyer.

When Edgar had called to offer me a position on the crew and informed me that I would be a sawyer, I felt a moment of elation. It is a great honor to be a hotshot sawyer. Sawyers are seen as the knights of the firefighting world, punching into brush to circle the enemy. They polish their saws, sharpen them, talk to them, and christen them. Some saws live in legend: "Tweaker Bitch" is dependable, "Gator Tail" rips, "Thrasher" had a good run.

Amid the laughter, my elation turned to dread. When cutting line, sawyers set the pace for the crew, so it is more than a job. It is a leadership position; one that experienced hotshots thought they deserved and I, a newcomer, did not.

"All the new guys think they're sawyers," someone said. A third year named Barba was staring me down. His arms were dark with tattoos. The chainsaw had not won me any friends.

I hadn't wanted to be a sawyer on the hotshot crew. Not really. I would have been perfectly happy carrying a hand tool, working from the back of the crew, free to absorb the wisdom of the fire whisperers as we hawked tobacco from mountaintops and watched firenadoes cross the land.

Instead, I had cursed myself. When Edgar heard that Aoki had broken a sweat hiking with me, he chose me for his own squad. Hotshot crews are typically divided into two squads, "Alpha" and "Bravo," comprised of eight men and two leaders. These squads are organized out of two short forest-green buses called "buggies," which are lifted and fortified to pound down rutted roads through flaming forests. The exteriors of the buggies unfold, revealing the tools of the trade:

chainsaws, axes, hoes, spades, flare guns, incendiary grenades, and drip torches.

Other firefighters call hotshot buggies prison buses because the interior is a dark den of stink and soot. Each hotshot occupies a three-by-three-foot space, called their "world," where they stash all their possessions for six months: rolling tobacco, chewing tobacco, snorting tobacco, canned oysters, canned sardines, powdered greens, surf magazines, bikini magazines, and sometimes a book or two. In a bin above their world, they stash a sleeping bag, three shirts, and an extra pair of underwear. For six months, the buggy would be my home, and the crew would be my universe.

"Okay," Aoki said. "Most of you know the drill. Everyone load up." The circle broke, hotshots hustling to the buggies. We were going to fight an imaginary fire.

In the buggies, each squad develops its own dynamics. As we climbed into the back hatch and rumbled out of the driveway, the returning hotshots offered the new guys a quick cultural orientation. Bravo Squad was technically run by a bear of a Dutchman named Verdries, who was middle aged, had a big beard, and liked to travel. But everyone knew that the shots on Bravo Squad were really called by Axel, a bright-eyed, thirty-five-year-old surfer dad. With over a decade as a hotshot, Axel was one of the most experienced members of the crew, a status that he carried with an ease that matched his love of Bob Dylan music and earthy spirituality. At night, Axel slept facing south to align himself with the electromagnetic currents of earth. Bravo Squad under Axel and Verdries seemed like a land of equality, laughter trickling from their buggy windows to torture the members of the austere Alpha Squad.

I sat in the back corner of Edgar's buggy, across the aisle from a twenty-five-year-old sawyer named Scheer, who was entering his third year on the crew. Scheer was an artist, surfer, and notorious party boy. He was skinny and tall, with an unruly mop of hair. He had a big nose, long neck, and wiry arms tattooed with eyeballs on the triceps. He looked like he had to concentrate to stop smiling. As the buggies tunneled down a forest road, Scheer told me in a hushed voice what to expect from Alpha Squad.

Edgar didn't like laughter in his buggy, or any noise at all that interfered with his death metal music. He enforced a rigid, militaristic hierarchy. Edgar enjoyed peering back at us from the passenger seat, expression inscrutable behind his sunglasses, before barking out some terrible insult over the screeching guitar. "Bravo Squad is the happy wholesome family all the kids are jealous of," Scheer confided. "Edgar's the abusive foster dad, and we're his adopted orphans."

"It's the happy families that always fall apart," observed Drogo, Scheer's puller, from the seat ahead. The hotshots had dubbed him Drogo—after the steroid-pumped Soviet boxer in Rocky Balboa films—to cement his reputation as a big dumb brute. At six foot five, 230 pounds, Drogo was an ideal puller because he often got so angry that he would rip trees from their roots before Scheer could cut them. With a chiseled jaw and brooding aura, he looked like a GQ model, but he also smoked cigarettes, rode a motorcycle, and carried a pistol—an identity he had worn like a costume, for amusement, ever since he trained with the Special Forces. When Drogo told me he had purchased his cop sunglasses at a Blue Lives Matter rally, he winked. The macho facade masked his New England liberal upbringing, a private high school education, and his sharp, playful mind.

The buggy jolted to a halt. Before the engine was off, Scheer, in a fluid motion, pulled on his fire pack and dove out the back door. I was right on his heels, pulling my chainsaw from the exterior compartment and hurrying to the trailhead, where we waited for the rest of the crew to file behind. There was one straggler, whom Aoki stared down until the disapproval was palpable. "Let's go," he finally said, and began walking.

Together, we followed the trail into the mountains. Fragrant sagebrush, manzanita, and chamise formed a ceiling of overhead brush. At some unseen signal we stopped. Before I could catch my breath, the men in front of me fired up their chainsaws to cut "practice line," a path of unburnable dirt around an imaginary fire. Scheer sliced into the green, opening a path that I expanded to six feet so the men behind could scrape a line with their shovels, axes, and hoes. After an hour fighting through the thicket of brush, the sun drawing moisture from my body faster than I could hydrate, we moved into a boulder field.

With the laughter of the hotshots still echoing in my skull, I had a paranoid sense that the crew was scrutinizing my skills, and I lost focus. My chainsaw teeth chipped a rock, throwing sparks and dulling my chain. I moved to cut the next bush. The chainsaw just spun against the wood. Scheer disappeared ahead. The hoes, shovels, and axes chinked closer behind. My shirt was soaked through with sweat, more from stress than exertion. I was slowing down the crew.

Hotshots know that falling behind on a hike or while cutting line is caused by a state of mind, a loss of focus. If you slowed down the crew in the wrong situation, the fire could trap you. I didn't want to be in that situation. Even more, I didn't want the crew to think I might put them in that situation. There was no quicker way to commit social suicide on a hotshot crew than falling behind.

I heard a shout and turned to find Edgar standing in front of the scrapers. His mouth moved, but I couldn't hear anything through my earplugs. His face transformed into a mask of rage. I read his lips—*Fuck you!*—and flinched at the shovel he hurled in my direction. I turned off the saw. In the silence, Edgar growled, "Cut like that on a fire and you're going to get us all killed."

◈

AT FIRST, I FOUND THE HYPERMASCULINITY OF THE HOTSHOT CREW jarring. I hadn't been part of such a competitive group of men for over a decade, since high school sports in the Midwest. I found myself digging deep into my memories, trying to regain an intuition of how to act among men—when to perform, when to stand up for myself, when to cower. The mental effort was exhausting, so our time in the classroom became my escape.

Some afternoons, after cutting line, we met in Aoki's office building, filing into a back room crowded with desks. The room reminded me of elementary school, with an old PC crammed into a corner and educational posters covering the walls. But on the posters, instead of animals and alphabets, were lists hotshots could memorize that would help keep them alive: the eighteen "watchout situations," the ten "firefighter orders," the "five common denominators" of firefighter deaths (fire running uphill, unexpected wind shift, light and flashy fuels, fire deceptively quiet, and critical burn period between 2:00 and 5:00 p.m.).

In the classroom, we were seated elbow to elbow, most men looking slightly pained. Besides me, only two held college degrees. The rest seemed traumatized by school. The windows were fogged from body heat. The air smelled like sweat. The only sounds were whispers and chuckles and the squelch of tobacco spit.

"People talk about how experienced we are," Aoki told us before our first lesson. "But the only real knowledge is the knowledge of the fire you're on. *That* is the experience that matters. All of this," he gestured toward the blank whiteboard, "all of this is to help you understand what you're seeing."

The experienced hotshots rotated around the podium, clicking through standard government slides that showed us how to identify and avoid the many hazards of the fireline: stinging bees, falling trees, and shifting fire patterns. But principle among the hazards we would face, we learned, was the danger posed by our own masculinity.

Groups of people, when alone together for long periods of time, often develop something like a collective chemical biome. Historically, male testosterone levels were presumed to be biologically innate, with genetics predisposing men toward aggressive, competitive, and sometimes violent behaviors. However, testosterone levels are also profoundly influenced by social contexts. In environments where individuals are expected to compete for power within social hierarchies, this pressure can cause testosterone levels to spike.

For hotshots, the superficial influence of this testosterone boost can be comedically rakish. A PowerPoint slide showed us multiple natural objects that, for men alone together in the woods, may come to resemble suggestive human body parts. From a certain perspective, a rocky protrusion appears phallic. A pair of hills becomes breasts. A split tree trunk looks like spread legs. On the PowerPoint slide, a red X was superimposed over these images. "Do NOT," the slide read, "use human anatomy to describe the fire environment."

We were cautioned to be aware of and mitigate this masculine energy—which the Forest Service called "machismo"—not simply because it is gross or inappropriate, or to avoid offending the women who work on wildfires, but because the social sphere of machismo can become terribly dangerous. Axel, the college-educated leader of Bravo Squad, seemed particularly attuned to this. He described multiple fires where people died, in large part because competitive pressure compelled men to take unnecessary risks. He cautioned us to beware of blind obedience, to always speak up if we felt like we were in danger, and to never, ever sacrifice ourselves for our mission. Scheer coughed in the back of the room, and his cough sounded a lot like

Edgar. Axel pretended not to hear. "Always ask yourself," Axel said, "'Am I doing this for pride, or because it's the best plan?'"

This was easier said than done. Even as we were implored to beware of hypermasculinity, we were engaged in an enterprise that, from its inception, has been built atop a foundation of hypermasculine values. For about a century, scholars imagined that men were naturally inclined to suppress fire. Carl Jung, a prominent Swiss psychologist in the twentieth century, believed that fire itself was a universal symbol of male characteristics, representing active, assertive, and transformative energy. Sigmund Freud took the symbol more literally. In his telling, fire suppression is driven by both male psychology and anatomy. There is "no doubt," he wrote, "that flames shooting upward were originally felt to have a phallic sense," prompting males everywhere to urinate on fire in a homoerotic duel with nature. Chainsaws and hoses, according to this line of thinking, eventually became extensions of male biology, allowing modern man to transform an evolutionary impulse into all-out war.

War was on the American mind in 1948, when the U.S. Forest Service founded the Los Padres Hotshots, among the first specialized wildfire suppression crews in the nation. Throughout World War II, fire had been a specter in the mountains, a domestic threat to the timber reserves fueling military campaigns abroad. "Careless Matches Aid the Axis!" screamed local headlines in 1941, beneath leering faces of Hirohito and Hitler. The same year, the *Los Angeles Times* warned of "the havoc wrought by the red demon," with the State Board of Forestry calling for a "firefighting army."

In the early twentieth century, wildfires were typically contained by local groups who filed into the mountains with spades and mule trains, but images of industrial warfare motivated the public to think bigger. "Bigger and stronger tanks, greater guns, larger battleships," wrote a 1941 opinion piece in the Santa Barbara *News Press*, "should develop in the imagination of some minds the idea of a national war against forest fires." The author called for the deployment of "firefighting equipment as gigantic and as powerful as any engine of destruction yet conceived."

While the American public no doubt imagined that industrial machinery would vanquish this red demon, the nation's new hotshot crews quickly gained a reputation as the most effective resource. Their mission was simple, outlined by a federal edict in 1935. Any fire prowling the American West should be corralled and exterminated by 10:00 a.m. the following morning. This was assumed to be a task for men because, as Freud wrote, throughout evolution, women's "anatomy made it impossible for them to yield to the temptation" to extinguish flames.

These sorts of intellectual acrobatics primarily function to make the macho values of fire suppression and the gendered hierarchies of society seem natural. In 1976, one of the first female hotshots in the United States began her career on the Los Padres Hotshots crew. This was supposed to be a watershed moment that would diversify the profession, but it proved a token. Almost half a century later, women make up about 12 percent of America's wildland firefighting force. Of those, less than 5 percent remain in the career long enough to gain leadership positions. A significant reason for this imbalance is the fact that, instead of entering a workforce where the values and practices have been reformed to accommodate and reward diverse approaches, women in the Forest Service are often judged by the same core machismo values that have persisted all along. "When you see a chick on a hotshot crew," a firefighter once told me, "you know she has bigger balls than you do."

With the hotshots, I began to feel that we were all trapped in a bind of cultural momentum. To survive on the crew, we needed to learn to embody the very traits that could kill us.

THE FIRST TIME EDGAR THREATENED TO KICK MY FACE IN, I KNEW IT was coming. It was the end of the first week. We were cutting practice line again, imagining a fire burning through an oak grove in the mountains. And, of course, I was falling behind. I could feel Edgar's eyes crawling down my back. I panicked. I tried to cut faster, but lost control and buried the tip of my saw in the dirt, dulling the chain.

So, when Edgar's hand clamped on my shoulder, and I turned to find his face inches from my own, smelling of spearmint gum, I was not surprised when he mimed a throat-slitting motion with decapitating force and told me to kill the engine.

I turned off the chainsaw. In the silence of the forest, Edgar's words weren't necessarily a threat, but more of an observation, growled loud enough for the rest of the squad to hear, that he *would* kick my face in, due to my amateur cutting style, if he were my puller. My puller, a young man named Smitty with a cherubic face and mercurial moods, unfortunately seemed to agree, fuming waist-deep in branches I had rained down on him in my attempt to catch up. The rest of the hotshots leaned on their tools, watching with amused expressions as Edgar twitched in rage.

"Your cuts *suck*!" Edgar yelled. He ripped the chainsaw from my hands and held it over a small tree I had dropped over our imaginary fireline. This, he demonstrated, was a case in point. Instead of cutting the tree from the base, to make the job easier for my puller, I should have sliced it into three pieces, starting from the top.

Then he pointed up to the canopy. "Does that look like six feet to you?" he asked. I genuinely couldn't tell. The fireline was supposed to be a six-foot gap in the forest, devoid of vegetation from the ground to the sky. But the trees were shifting in the wind, making the sliver of sky move. Edgar thrust his chin forward. "Does it?"

"I guess not."

Edgar shook his head in disgust. He told the rest of the squad that we were finished for the day, so they should clean up the fireline before Aoki came to inspect it. Then he came closer to my ear. "You're not ready yet," he whispered.

Those words stung more than the shouts, because the whisper wasn't a performance. He meant what he said. As I hefted my saw over my shoulder and followed the fireline out of the trees, the whisper sat in me like a weight. Edgar, I worried, was simply articulating what the rest of the crew muttered among themselves.

Throughout the week, the tension had grown like an ambient background noise, imperceptible at first, but rising slowly to a deafening pitch. The day before, at the station, as I went to wash my dishes after lunch, I had passed a group of men who stopped talking and stared until I was out of sight. At the sink, I bumped into Barba, the tattooed Marine veteran who had glared at me during introductions. To smooth things over, I asked Barba how I was doing.

"You don't wanna know, bro," he said, looking down at his dishes. "You don't want to know what people are saying about you."

"Man, it's that bad?"

"Look," he said, setting down his bowl. "I know you don't make the rules, and if the overhead wants to break tradition by having a new guy on the chainsaw, that's their decision. So what if there's people, like me, who've been on the crew for three years? They should at least make sure the new guy can *cut* before giving him the saw. It's on them, man. Not you." He slapped me on the shoulder, then walked away.

After Edgar berated me, I joined the other sawyers in a clearing in the noonday heat. I didn't talk to anyone, but dropped to my knees to sharpen my chain. I wanted them to see that I was taking the job seriously. Thighs straddling the body of the saw, face close to the teeth, I drew a file along each crescent hook, listening to the rasp of metal, feeling the kinks grow smooth.

"What in hell are you wearin', Thomas?"

I looked up, squinting through my safety glasses, which I had donned in a conscious attempt to avoid getting yelled at. The fine metal particles scraped from the chainsaw teeth while sharpening can get into your eyes, causing permanent damage to your vision. Axel, the leader of Bravo Squad, had advised me to always wear glasses when sharpening the saw.

Now Tyler, the lead sawyer, stood over me. He was a unique breed of mountain man, mid-thirties, short, hairy, and strong. Scheer came to stand beside him. In the offseason, they surfed together in Indonesia. Since it was Scheer's third year on the crew, he was supposed to be demonstrating leadership, so he crossed his arms and looked down at me in disapproval.

"Lose those glasses, Thomas," Scheer said in his surfer voice. "You look like a total kook." Scheer had recently told me that a kook was the pinnacle of hotshot insults. Call me a wimp, call me a jerk, but never a kook. A kook was a tech bro with a Tacoma truck, a city slicker who needs rescuing on a day hike, a surfer caught in a riptide. Or a rookie sawyer who wore safety glasses to sharpen his saw.

"He looks like a chemist," cackled the lead sawyer, looking down at me. "A goddamn *mad scientist*. An embarrassment! Take off those glasses, Thomas."

Red, attracted by the attention, came to join the others standing over me. He was Edgar's assistant on Alpha Squad, a thirty-year-old redhead of Irish descent with a lip full of tobacco. He furrowed his brow, as if he felt Edgar watching. "You know, you really shouldn't be field sharpening your saw until you get some more experience, Thomas. Wait until we're back at the station. You could really damage the chain."

I stayed quiet. The saw, I was coming to realize, wasn't the point.

Every interaction during training was a performance, an attempt to mold the hierarchies of the crew.

Hotshots pride themselves on their lack of rigid hierarchy. "Hotshots don't do all that 'yes sir, no sir bullshit,'" a retired hotshot told me. "That's for Cal Fire. Our respect is founded on experience. It's earned." Instead of rules, the hotshots regulate respect informally, using gossip, cold shoulders, and derisive jokes to keep one another in check.

The system, stressful as it was, made sense to me. On the fireline, margins are so small and errors so potentially catastrophic that every hotshot has the authority to reprimand others. And because catastrophes could hinge on the most absurd details, there was virtually nothing in the daily routine that fell outside the group's purview. Whether you tied your boots or drank enough water or sharpened your chainsaw were all matters of public concern, and were open to scrutiny by everyone.

The whispers, gossip, and jokes forced an attention to detail, a constant clarity. By trying to gain their respect, I was being disciplined into a new way of thinking in which I scrutinized each of my actions before it could be scrutinized by others. They were dissecting my mistakes because everything I did as a sawyer could have serious consequences for the rest of the crew.

I knew all this in theory. But it did nothing to assuage the sensation that I was stuck in a downward spiral, close to being crushed by collective disapproval. "We're cannibals," Scheer told me. "Do your job right, and everyone supports you. Step out of line, we eat you."

❦

ONE EVENING, WE STOOD SHIRTLESS DRINKING COORS AROUND PICKUP trucks under the shadow of an oak tree. The new guys gravitated toward each other, so I found myself standing by Jackson. Jack was stout, with canny eyes, an easy laugh, and the largest beard on the crew. He had worked as a hotshot several years before, in Arizona, but quit when they told him he needed to trim his beard. It seemed he hadn't trimmed it since. He joined the Los Padres Hotshots to be closer to his family, who lived in San Diego, so they could help care for his two-year-old daughter while he was away.

Jack was several beers deep but wasn't concerned about the drive home. He had parked his trailer right at the edge of the hotshot compound. "My old lady has dinner on," Jack said with foam in his mustache, "but I told her this is part of the job."

I found myself slightly envious of his carefree attitude, which I speculated aloud must be because he was on Bravo Squad. Jack replied that the grass is always greener on the other side. Bravo Squad *seemed* nicer, he said, but it was like living with a bunch of girls. They never said anything to his face but were terribly passive-aggressive. Jack smoked cigarettes, drank lots of beer, and, in his midthirties, was older than most on the crew, so he had a hard time keeping up on hikes. Rather than berating him as Edgar would have on Alpha Squad, someone in the Bravo buggy had drawn an unflattering cartoon of Jack with a speech bubble saying, "I don't like to work." It was posted where everyone could see it.

Regardless, I thought it must help Jack's case that Axel had taken to him with the enthusiastic amusement of a child with a new toy. There's no other job in the world, Axel told me, where such an odd cast of characters are forced to live in the woods together for six months, like an uncontrolled social experiment with more conflict and comedy than any reality TV. Jack called himself a hillbilly, but Axel called him a critter, a creature, a man-thing belonging to a different time or another world. As a teenager, before ancestry tests began debunking wishful origin stories, Jack thought himself a direct descendant of a racist Civil War general and tattooed his forearms accordingly: "Confederate" (left), "Royalty" (right). At thirty-four,

Jack didn't believe himself a racist because he bonded with conspiracists of all colors, especially those who could connect the dots between COVID, Bitcoin, and Joe Biden: "I know it sounds crazy," Jack said, "but it all makes sense."

In the moment, I didn't care how Jack made sense of the world. I was just thankful he hadn't mentioned my troubles on the chainsaw. When Jack wandered off in search of another beer, Scheer caught my eye and pulled me aside.

"Be careful with Edgar," Scheer whispered. "He's the antichrist." Edgar had terrorized Scheer since he joined the crew three years ago. "My mom's my therapist," Scheer said, "and she hates the guy." Edgar and Aoki were often at odds politically but trusted each other to keep the crew alive. "The difference is, Edgar likes to humiliate people. Big MAGA guy. He thinks yoga is gay. Doesn't believe in climate change."

"Nah," Márlon glided over, sensing gossip. After five years on the hotshot crew, Márlon still moved with a dancer's grace. He immigrated to the United States from Nicaragua as a teenager and looked like a twenty-nine-year-old Antonio Banderas, with a stern face but mischievous eyes. "Edgar's nice this year," Márlon said. "He has a girlfriend now. This is happy Edgar."

Scheer agreed with that. His first year on the crew, fighting a fire in Alaska, Scheer had asked Edgar if he was scared, trying to make a friendly joke. Edgar had marched over and made an exhibition of berating Scheer for ten minutes straight, spittle flying everywhere. He then ordered Scheer to clear a log-blocked road alone. For an hour, Scheer bucked fallen trees while Edgar and Alpha Squad crept behind him in the buggy, hooting from the windows.

"He told me I would never be a sawyer," Scheer said. "On my saw certificate, he wrote 'bucking only.'" This meant that Scheer was only allowed to chop up fallen logs. "But the joke's on him." Now that

Scheer was the second saw, a prestigious position, he regularly dropped big trees. He had named his new saw "Bucking Only" in a backhand slight to Edgar.

"It's tough love, homie," Márlon told me, clinking my beer. "It's how he shows you he cares about you. It'll make you a better hotshot."

"He's just old-fashioned," Red agreed, joining the conversation. Red was one of the only hotshots wearing a shirt. He seemed self-conscious of the belly he had accumulated during his past year recovering from a torn ACL. But he was tough, could work nonstop for days without sleep, and had been promoted this year. Coors in hand, leaning against the tailgate of a Tacoma in the dappled light, Red seemed at ease. He was a kindhearted, auburn-haired country boy who enjoyed the small things in life, a respite from the barking pressure Edgar imposed on him throughout the day.

"That's how it used to be," Red continued. "Edgar's been a hotshot since he was eighteen. Things were different back then. When he joined, new guys got roasted for years. He doesn't want people to get soft."

"I hear he had it rough," Márlon added.

This was a rare sentiment from Márlon. Among the hotshots, Márlon had gained a reputation for an inhuman tolerance for danger and pain. Márlon was Red's right-hand man on Alpha Squad, Red told me, because he understood wildfire and would keep the boys working all day. But he hadn't been promoted for several years because Aoki didn't trust him not to get the boys killed.

The sun dipped behind the mountains, pulling shadows across the golden slopes. One of the hotshots was demonstrating how quickly he could chug a beer. Márlon and Red wandered over to cheer him on.

Scheer leaned in. "It is different now, you know," he said. "When I joined, we were treated like shit for the whole first year. That's whack. That's not the vibe we want here. We're changing things."

I had arrived at training half expecting to get stuffed into a dumpster, so I told myself I could handle a few aggressive men. But as I drove home over the mountain pass, city lights glowing like embers against the ocean, my emotions felt frayed.

That night, after eating a bowl of cold pasta Kenzie left out for me, which I was too tired to microwave, I slept restlessly. Edgar haunted my dreams. I was frozen amid the brush, fire rushing toward me, chainsaw useless. Edgar's face filled my vision, twisted in a silent howl.

"YOU'RE OFF THE SAW, THOMAS. YOU'RE PULLING TODAY." EDGAR TOLD me this the next morning, in passing, as we loaded into the buggies to practice cutting line.

"Copy." I knew it was meant as a demotion, but I was flooded with relief. Finally, I would be able to occupy my proper place as a new guy, hopefully mending my relationship with the crew.

Márlon came to stand beside me. He leaned against the buggy with a plastic cup of coffee. "Ah, Thomas, you messed up, homie." His face was set in mock concern, but his eyes were laughing. "You pissed off Edgar."

"Did I?"

"Nah, I'm just playing. You're good."

I felt good as we hiked back up the trail in the hot sun. It wasn't just the weight of the saw lifted from my shoulders. It was the knowledge that I wouldn't be setting the pace for the crew, that my life would be simple now. The crushing scrutiny, the sense that I had jumped my proper status, all that was gone. Now, I just needed to throw brush.

I practically floated up the mountain, enjoying the sound of marching boots and creaking leather, the breeze cooling my sweat and rustling the leaves. I watched Smitty lumbering in front of me. He was a second year who deserved the saw that was slung over his shoulder. That chainsaw is social poison, I thought. The six-month fire season, which before had stretched in my imagination like an infinite hell, finally seemed possible. I might even be able to make friends.

Then Smitty crumpled.

He collapsed without warning, like he had been shot. One moment, he was marching with my chainsaw. The next, he was on the ground, body limp, arms and legs splayed, eyes rolled back, lips quivering.

"Medic!" I shouted, after a breath. This was the only word I could think of. "Medic!"

Aoki took a few more steps before slowing to look back. "What?" He looked annoyed, then confused when he saw the body on the ground. "Are you serious?"

"Medic!" I repeated, panicked, staring at Smitty. His face had grown pale, hands clawing, muscles curling in on themselves like he was entering a seizure.

Axel appeared from nowhere, blue eyes hard, giving orders in a curt, calm voice. I stepped back to make room as Axel tore open Smitty's shirt. Everyone was yelling, moving, pulling out radios and cell phones to try to call for help, but I couldn't make sense of any of it. I backed away with a ringing in my ears.

I looked over the valley, away from the chaos, my eyes following webs of forest that branched up ravines and gullies over the mountains, toward my home. I thought of my abandoned office at the university. It seemed absurd, impossible somehow, that by crossing the mountain pass, by joining the hotshots, I had entered a world where a colleague could lie contorted at my feet, surrounded by frightened faces, fingers pressed to his neck for a pulse.

Edgar was pacing, moving with a dark energy through the crowd. He noticed me on the periphery and stormed over.

"Is this your pack, Thomas?" His eyes were wild, his voice dangerous, bringing me back to the present.

I had no idea what he was talking about. Everyone had dropped their packs, and each pack looked the same.

"Answer me!" He held up an empty pack, which he had mistaken as mine. I realized he thought Smitty had been carrying my extra weight. I was overwhelmed, speechless. He interpreted my silence as guilt.

"This is on you!" he screamed, pointing at Smitty's body. "This is on *you*!"

After Edgar stalked away, Márlon came to stand beside me. He was calm, as always. I sensed that he had been in such situations before.

"Well, homie, looks like you're back on the saw."

CHAPTER 4

AFTER OUR FIRST WEEK OF TRAINING WE HAD TWO DAYS OFF. Aoki told us to enjoy them. We would be one of the first hotshot crews in the nation to appear on the federal resource list as "fire ready." Fires were already burning in the Great Basin, so after our days off we should expect to roll nonstop for the rest of the season. The first day, I woke to a text from Scheer inviting me to play volleyball on the beach. I was honored to get the invite, but I deferred. I needed a break from the testosterone.

It was late morning by the time I woke up. I found Kenzie as I would any other day, seated on the floor in her pajamas with her back against our couch, typing away on her laptop. She was working on her own PhD, researching sustainable food systems. Sunlight flooded the east-facing windows and made our dozens of houseplants glow. Though I had been at training for only a few days, the room felt different to me, as if the stress and excitement and exertion of those days had slightly shifted the tenor of my belonging here. When Kenzie noticed me staring at her, she cocked her head and smiled, the same as any other day. In her routine, my time at training had slipped by. But for me, this felt like the last gasp of a life I was about to leave behind.

"Good morning," she said.

"Good morning."

Kenzie asked if I wanted to go for a jog.

No. Recreational exercise sounded dreadful.

Did I want to go to the beach?

No. I'd prefer to avoid the heat while I could.

Did I want to grab lunch?

No. I was spending too much money on beer and coffee for the crew, and I hadn't yet received my first paycheck.

I was out of sorts, and I didn't know why. People often view identity as something we project outward to others, like our core possesses an innate being. In reality, most anthropologists argue, our identity hovers in that relational space between ourselves and those around us. After just a single week, I was finding it difficult to shift gears between my two lives. Kenzie, as if sensing this, tried bringing me home. She insisted that we walk our dog down a footpath to a nearby park to pick oranges for juice. I agreed.

The town was in full bloom, a flood of colors and scents, with lavender lining the streets, bougainvillea and wisteria draped over homes, and jacarandas shedding blossoms in a purple rain. As we walked, I asked Kenzie about her week, but I wasn't really listening. I was going through the motions; my thoughts were drawn instead toward practical matters, the nuts and bolts of chainsaw maintenance, the protocols I was expected to memorize, and the propane canisters I needed to purchase for the crew, on my own dime.

After filling a bag with oranges, we set out to harvest a salad for lunch, following a foraging map Kenzie created to find miner's lettuce, nasturtium, and wild mustard. She liked to say there's no such thing as a weed, just plants we're not yet familiar with. We brought our salads with us as we rode our bicycles toward the university. Our dog followed, paws pattering over seaside cliffs where college students read books and tanned in the sun.

We stopped at a dog park to look out at the Channel Islands some thirty miles offshore. On their windswept shores, archaeologists discovered the oldest human remains in the Americas, dated to some thirteen thousand years ago. When those first people traveled down the California coast, they would have been familiar with fire's use in landscapes. From their journeys, and the ancestral knowledge they carried with them, they would have known that herbivores congregate in burned areas to browse fresh shoots. Before long, through careful observation and experimentation, they would also have learned the nuances of this specific place—the medicinal properties of plants, the edibility of the local flora, and the variable impacts of different kinds of fire on the biome. In the archaeological record, when humans first arrived here, evidence of flames in the landscape blossomed alongside, growing from small areas burned by occasional lightning strikes to permeate every place these people made their home.

By the fourth century BCE, the descendants of the first people had established a thriving civilization along the Santa Barbara coast. They called themselves, as they do today, the Chumash, which translates roughly as "seashell people." In the years before the Spanish came, around 150 Chumash towns dotted the foothills and plains between the ocean and mountains, the largest of which was located a few miles from my apartment. While the people inhabiting inland towns hunted, cultivated protein-rich grasses, and gathered acorns, the coastal settlements became maritime trading ports where people harvested surfgrass, speared seals, and wayfared into the sea in pursuit of swordfish. By creating elaborate networks of exchange that distributed the land's biological diversity among the different towns, the Chumash enjoyed a biome that was among the most abundant in the Americas.

All of this was made possible by the intentional use of fire, even for those who made their living from the ocean. Soaproot, an innocuous native plant easily mistaken for a weed, is a case in point. Soaproot grows best in the open conditions created by intentional burning, and it flowers when stimulated by fire. The Chumash used fire to encourage soaproot because, they discovered, when the bulbs are placed in a stone oven and baked for half a day, they become a source of nutritious food. When pounded and sprinkled into water, the bulbs would also temporarily paralyze fish, causing them to float to the surface so they could easily be collected. The same pulp can be used as soap for washing hair, skin, and clothing. When dried, it can be made into a resinous coating that increases the longevity of tool handles, baskets, and building materials. To understand the centrality of fire to the people and ecology of this region, take the many uses of this single plant and expand it to include the roughly two hundred other fire-loving species that inhabit the coast.

The fire-facilitated biological diversity of this region afforded the Chumash flexibility to develop expertise in boatmaking, basketry, and arts. They also enjoyed a great deal of leisure time, which they filled with a sport similar to soccer, and footraces from one town to the next. The most festive annual occasion was the winter solstice, which culminated in the "Day of the New Sun." Each year, people would travel from across the region, converging on whichever town was selected to host the event. They welcomed the new year with days of prayer, ceremonies, and political discussions, enlivened by singing, dancing, and sports. When the social and natural worlds were in balance, the fires of the region remained healthy, acting as elemental conduits of this abundance. But when fires became destructive, they signaled an imbalance between people and the land, and people with each other, so the Chumash would use these gatherings as occasions to identify the sources of discord and rectify the relationships at their root.

Standing on that cliff with Kenzie, looking at the island that appeared to float in the sky, I felt caught between my professional duty to ensure I would be equipped to suppress fires when the time came and an intellectual urge to understand where the old fires had gone and what had set these relationships awry. While Santa Barbara was where the first people of the Americas brought fire to these landscapes, it is also the place where the culture of fire suppression first took root.

To join the hotshots, I had taken a leave of absence from my academic post, but I still had access to the archives. I suggested to Kenzie that we spend the rest of my days off catching up on research. I tried framing this as a couple's activity, but she knew better. She gave me a long look and shrugged, which I interpreted as approval. I requested a stack of old documents from the archives in Santa Barbara's Spanish mission, and I settled in.

IN THE SUMMER OF 1769, THE FRANCISCAN FRIAR SHAMBLING ACROSS the Sonoran Desert was an unlikely figure to bring fire suppression to California. Stooped at less than five feet tall, Junípero Serra walked with a pronounced limp. His leg was ulcerous, possibly from a snakebite in the Sierra Gorda mountains of Mexico, but Serra relished the pain. As a Spanish Inquisitor, Serra's medieval brand of piety demanded a ferocious self-torture: he slept on a board, whipped himself as he walked, and was known to burn and lacerate his own flesh during sermons. Beneath his gray friar's smock, he wore a sackcloth spiked with bristles. When he arrived in California, Serra's enthusiasm for corporal punishment would allow him to complete the long-awaited

Spanish task of subjugating the Indigenous people of the coast, stamping out their flames and opening their land for extraction.

Over the past 227 years, California had gained a reputation among the Spanish as a shadowland of rocks and fog where expeditions went to die. In 1542, the first attempt at exploration was fanciful—a classic conquistador quest to find a lost city of gold—commissioned soon after Hernán Cortés had conquered the Aztec Empire and founded New Spain in present-day Mexico. After nearly a year at sea, however, the expeditionary ships limped south again with a demoralized crew and little to report, having buried Cabrillo, their commander, on a windswept island off the coast of Santa Barbara.

Later expeditions fared worse. In 1584, a ship commanded by Francisco de Gali discovered that the quickest trade route from the Philippines to Mexico followed a current across the Pacific Ocean to California's north coast. Within several years, this route became the principal connection between Asia and North America. The journey, however, took as long as two hundred days to complete. Anyone perched on California's cliffs observing the passing ships would have witnessed a parade of galleons crewed by half-dead, flea-ridden sailors with atrophied bones, bleeding gums, and festering sores. Naturally, the viceroy of New Spain commissioned a ship to scout the California coast for a resupply port, but in 1595, this expedition wrecked near Point Reyes. For generations after, the Coast Miwok tribe would fashion ornaments from the officers' porcelain that washed ashore.

By 1602, after just one more feeble attempt to chart the coast (which resulted in a map so infused with the captain's vainglorious exaggerations that even the conquistadors didn't trust it), the Spanish all but abandoned the conquest of California. For the next 167 years, the land was relegated to the hazy edges of empire, the brink of the known world, where expeditions were doomed.

Jesuit priests stepped in to fill the void of European governance.

By the end of the seventeenth century, Jesuits had settled the northern edges of the Spanish Empire throughout Sonora and the Baja Peninsula. While these priests were a vanguard of European expansion, they also served as a buffer against the brutal Spanish settlement practices that had proven so catastrophic for Indigenous people across the continent. The Jesuits believed that Catholicism should adapt to the cultures being evangelized, so their priests often embedded themselves in Indigenous communities, learning the languages and adopting local customs. By the middle of the eighteenth century, the frontier of New Spain was peppered with communities that blended Jesuit and Indigenous practices, where priests used their political clout to repel intrusions from fortune-seeking civilians and bands of marauding conquistadors.

For those seeking to profit from the "untapped" resources of California, however, the buffer zone of Jesuit priests became a problem. By the 1760s, as steam engines began to chug across England and tipped Europe into the Industrial Revolution, the old Spanish dreams of discovering cities of gold transformed into more practical goals of generating fiscal value from their territories. It was no longer enough to claim land—that land must be exploited and settled, its resources extracted and sold. To clear the way for this new colonialism, King Carlos III ordered the expulsion of all Jesuits from the Spanish Empire. On June 24, 1767, the viceroy of New Spain reportedly read this royal decree aloud to the archbishop in Mexico City: "If so much as one Jesuit of this district, even though he be sick or dying, fail to be placed on the vessel [for expulsion], you shall suffer the death penalty."

And so, after over two centuries of halfhearted attempts to settle California, the Spanish Crown sent Franciscan friar Junípero Serra into the hinterlands of the empire to replace the Jesuits and to oversee the expansion of European civilization. With Serra, the first trickle of a new economic current arrived on the scarred back of medieval brutality and, with it, a new era of flames.

While most of this expedition—*La Sagrada Expedición*—set sail, Serra walked north from Loreto in the Baja Peninsula. He enjoyed the discomfort. But that July, when Serra crested the final pine-feathered coastal ridge and beheld the Bay of San Diego, he found the others languishing on the shore. Of the three ships that had departed San Blas months before, one sank, and over half of the three hundred men had died of disease and dysentery. The sheer quantity of death and suffering ultimately drove the expedition's only surgeon mad.

Undeterred, just over two weeks after his arrival, Serra founded the first Spanish mission of California. Initially, it was no more than a log shelter with a thatched roof, but it provided a base of operations for the series of twenty-one missions that would eventually line the coast, a day's journey apart, along the seven hundred miles from San Diego to Sonoma. The goal was to convert California's Indigenous people into Christian Spanish subjects who generated profit for the Crown.

In the early days of the missions, some Indigenous people came willingly, likely out of curiosity. "Every day Indians are coming in" from their homes in the mountains, Serra wrote in a letter. It brought him joy, he continued, "seeing about a hundred young children praying and answering individually all the questions asked on Christian doctrine, hearing them sing, seeing them going about clothed in cotton and woolen garments . . ."

Correspondence among the mission personnel, however, suggests that Serra's assessment was tinted with a rosy religious hue. As one soldier wrote from Monterey, "We had to live on rats, coyotes, vipers, crows, and generally every creature that moved on the earth, except beetles, to keep from starvation. How many times we wished we were six feet underground." For soldiers, the California missions became a frontier of isolation and insanity, a desolate corner of earth where authorities in Mexico exiled criminals to live out their days.

If conditions were dire for the soldiers, they were far worse for California's Indigenous people. Once Indigenous families were baptized at the missions, they were never allowed to leave without Spanish permission. Children were taken from their parents and confined to separate quarters where friars worked to eradicate their cultural traditions, which Serra considered pagan.

For Indigenous adults, the missions functioned as forced-labor camps. Men and women were segregated, and any activity that infracted Serra's strict religious dogma was punished with public floggings. Serra considered all Indigenous people to be spiritual children, so it was natural, he explained to the California governor, that they should be beaten. "Two or three whippings on different days," he rationalized, "may serve for a warning, and be of spiritual benefit to all."

The brutality of the mission system was so acute that it surprised even some of the Franciscan friars, who were accustomed to physical mortification. In 1798, a friar at Mission San Miguel wrote to the viceroy, "The manner in which the Indians are treated is by far more cruel than anything I have ever read about. For any reason, however insignificant it may be, they are severely and cruelly whipped, placed in shackles, or put in the stocks for days on end without even a drop of water."

The problem, then, for the Spanish, was to convince thriving Indigenous communities to abandon their own societies to endure the sadistic deprivations of the Spanish missions. Before the arrival of Europeans, starvation and food insecurity in the region were mostly unheard of. To force Indigenous people to the missions, and to harness the forced labor that would become the backbone of the budding market economy, the Spanish needed to identify and criminalize the foundations of Indigenous economic life. Within months of their arrival, the Spanish gaze turned to fire.

On an overland scouting mission, Friar Juan Crespí, a sallow-skinned student of Junípero Serra, began to notice signs of ubiquitous fire use throughout the landscape. Passing through the terrain that would become Santa Barbara, he entered a forest of oak and sycamore trees, the largest he had ever seen, with trunks charred at their bases. The hills between the mountains and sea were full of roses and chia grasses, "burnt in some spots and not in others," he recorded, "the unburned grass so tall that it topped us on horseback by a yard." As he continued north toward Monterey, signs of fire increasingly captured his attention. Camped atop cliffs near a large Chumash village, he complained in his diary that there was no grazing for the horses, because everywhere he looked was "burnt off by the heathens."

Several years later, the Spanish military governor would make this same complaint as he scouted Chumash territory along the Santa Barbara coastline. He recorded that he had difficulty finding places to stay overnight, or even sometimes places to rest at midday, "due to the horses and mules not having grass." The lack of grazing, he continued, was caused "by the great fires of the gentiles, who, not having care for more than their own bellies, burn the fields as soon as they gather up the seeds, and that [burning] is universal."

The governor was correct that burning was universal. Fire infused every part of California, from Indigenous economies, to the composition of soil, to the genetic makeup of plant species, the flow of water, and migratory patterns of animals. It is likely the Spanish interpreted fire as destructive because it enabled Indigenous people to survive apart from the missions, and therefore was not conducive to the economic profit they were seeking to create.

In the years that followed, Spanish complaints of Indigenous burning reached a fever pitch. By 1793, José Joaquin de Arrillaga, interim governor of the province of Las Californias, sat in Santa Barbara's presidio to review the complaints. He found that "the aforesaid damage is true." On May 31, Arrillaga penned the first official fire regulation of the American West.

The "childish" burning of fields, customary among Indigenous communities, was causing "widespread damage," Arrillaga concurred. "I see myself required to have the foresight to prohibit for the future all kinds of burning." He directed his administrators to "uproot this very harmful practice of setting fire to pasture lands," and to use "the most severe punishment" to implement this policy.

The friars strictly enforced Arrillaga's fire ban. A government questionnaire sent to the missions in 1798 sought to confirm that "the fathers have shackles, chains, stocks, and lockups" to punish misdeeds. A friar in northern Chumash territory replied that, for "transgressions against the common good, like . . . firing pastures," he opted to notify the military. By banning fire, the Spanish effectively criminalized the economic system of California's Indigenous people, justifying the use of force to bring them into the emerging ranching and plantation economy of the Spanish Empire.

This was a brutal transition. With their food systems criminalized, many starved. One man suffered five whippings for crying when his wife and child died, likely of malnutrition. Many were tortured and executed after trying to escape. Sexual abuse of women and children was rampant; one Chumash man penned a letter to Junípero Serra, reporting that his priest would, at night, enter the women's dormitory and force them to sing hymns, drowning out all other sounds. But Serra was already aware. "The soldiers," he wrote, "clever as they are at lassoing cows and mules, would catch an Indian woman with their lassos to become prey for their unbridled lust. At times some Indian men would try to defend their wives, only to be shot down with bullets."

Within just fifty years of the founding of the first California mission, by decimating the Indigenous people, the Spanish had all but eliminated fire from the landscapes of southern coastal California. Those fires that weren't suppressed with whips and shackles were extinguished by the toll of disease. Cut off from their homelands, prohibited from their cultural practices, and exposed to European viruses, some 20 percent of Indigenous adults died in the missions each year, according to Spanish records. The mortality rate for children was even higher—in some missions, 90 percent of children died before their tenth year. By the early nineteenth century, the Spanish mission system had killed approximately half of California's Indigenous people.

Faced with such carnage, some Indigenous people harnessed fire for resistance. Within months of the completion of Junípero Serra's first mission, the Tipai people of San Diego burned it to the ground. Six years later, the Tipai revolted again, killing the resident friar. In the decades that followed, the Chumash of Santa Barbara also revolted, while the Yuma people occupying a significant overland route to California began defending their territory from soldiers and colonists. By the turn of the nineteenth century, the flames of California's land had transformed into those of violent conflict.

Junípero Serra, for his part, approved of the violence. Seated in austere quarters beside the iron spikes he used to whip himself for sinful thoughts, Serra reportedly reflected on the broiling conflict. "Thank God that the ground has been watered [with blood]," he said. "Now, certainly, we will achieve the conversion of the Dieguenos."

This final point was a half-truth. Although Serra and his colonial heirs would never achieve their goal of erasing Indigenous cultures from California, they *would* succeed in a conversion of a different kind. By oppressing the people of California and the flames they carried, the Spanish sparked a conversion of the land, and the beginning of the megafires of today.

I EMERGED FROM THE ARCHIVES OF SANTA BARBARA'S SPANISH MIS-
sion and rode my bicycle home through a purple twilight. For the first time, I noticed the names of the streets that I followed: *Junípero, Serra, Arrillaga,* and *Crespí.* In 2015, when Pope Francis canonized Father Junípero Serra as a Catholic saint, he declared that Serra was "one of the founding fathers of the United States." Though misguided, the pope may not have been wrong in that declaration. The names inscribed on the streets of Santa Barbara hinted at the deeper ways these actions had inscribed themselves into both our landscapes and our culture, offering a window into the connections between the origins of fire suppression, the expansion of our nation, and the flow of historical violence into the turbulent conditions of today.

But the Spanish were not alone in their hatred of flames. They were, rather, caught in a global wave of fire suppression that, at this particular historical moment, Europeans carried to every habitable continent. In 1749, the same year Junípero Serra arrived in the Americas, Pennsylvania passed its first fire ban. A few years later, the New England colonies restricted fire "as a check upon that very destructive practice taken from the Indians." During that same period, Dutch trading corporations began executing Indigenous South Africans for burning the land. Meanwhile, British officials lamented that "the wild tribes" of India were "devastating the forests," leaving "a heap of ashes and irreversible ruin in their wake." As French industrialists colonized Vietnam, Laos, and Cambodia, they complained about "the fires started by the natives," which were "the plague of Indochinese forests." Even in Ireland, where farmers and shepherds had tended the land with fire since the end of the last Ice Age, the English, after invading, passed a 1743 law "to prevent the pernicious practice of burning land." Everywhere European colonizers laid their claims, they extinguished fire.

There was nothing inevitable about a European fear of flames. Prior to the development of complex industrial machinery and the acceleration of global trade, the common people of Europe used fire much as communities did around the world. In nearly every European environment, people burned to improve the growth of berries and mushrooms, to enrich pasturage and range, and to fertilize soils across heath, forest, and moor. Like in California, fire was an essential tool that communities skillfully used to make a living from the land. Fire suppression did not become commonplace until flames became a barrier to the economic interests of Europe's rising class of commercial elites.

The use of fire was prohibited in Europe at roughly the same time as it was criminalized in colonies around the world. In 1749, for example, a Swedish botanist noticed how useful fires were for the people of Scandinavia. His name was Carl Linnaeus. The son of a Protestant preacher, Linnaeus had achieved fame just over a decade earlier with his publication of *Systema Naturae*, a book that produced the modern scientific method for categorizing species. *Homo sapiens*, the name Linnaeus chose for humans, derived from the Latin word *sapere*—"to know." To know the difference between good and evil, Linnaeus believed, was what separated the human species from the beasts of nature. And what was "good" at the dawn of the Industrial Revolution was the codification, consolidation, and reorganization of the world into coherent parts that could be managed and extracted for profit.

For precisely this purpose—to make the lands of Sweden profitable—Baron Carl Harleman, a patron of science, sought out Linnaeus. He was not difficult to find. Linnaeus had recently returned from a botanical expedition to the Arctic, where he had exchanged his powdered wig for an elaborate costume of reindeer fur, magic beads, and a pointy woven hat. Linnaeus was frequently seen gallivanting in this outfit down cobbled streets, courting the woman he would eventually marry. But Baron Harleman interrupted this courtship with a request from the king himself. King Frederick I, Baron Harleman reported, needed Linnaeus to use his scientific prowess to produce information that would allow the government to transform rural farming communities into profitable industries.

Linnaeus was an indiscriminate documentarian. As he traveled from Upsala toward Skane through groves of whispering beech and towering pine, he noticed evidence of fire all around him. Although Sweden had recently criminalized the use of flames, Linnaeus began conversing with farmers and recording their burn practices. Their fires, he noted, were as varied as Sweden's flowers. Between forests, marshes, and peatlands, farmers used fire to release nutrients, charge soil with minerals, and prime grains, grasses, and root crops for growth. Fire, Linnaeus wrote, allowed the farmers to "get an abundance of grain from otherwise quite worthless land."

By the time he had examined Småland, a forested region, Linnaeus was thoroughly convinced of the importance of fire for small farmers. Their flames, he believed, were the only things preventing the land from regressing into a wasteland of rocks and sand. If the government were to strengthen their enforcement of fire bans, Linnaeus warned, the inhabitants would "want for bread and be left with an empty stomach."

But several months later, when Baron Harleman, the Swedish patron of science, sat in his office in Stockholm to read Linnaeus's report, he was enraged. Not only had Linnaeus "not condemned" burning by small farmers, a practice "so pernicious for the country," but he had "justified" and "sanctioned" fire use, which was illegal. For Harleman, Linnaeus's report was nearly unforgivable. He withheld further funds for Linnaeus's research until he removed all favorable portrayals of fire from the report.

What Linnaeus hadn't anticipated was that the "empty stomachs" he warned would follow fire suppression were part of the point. The criminalization of fire was one cog in a larger legislative machine. In France, Spain, Germany, and every other nation touched by the Industrial Revolution, commercial elites were in the process of criminalizing every tool common people had that would allow them to exist outside the system of wage labor. Governments banned hunting, fishing, and gathering grain in fields after harvests. They closed public lands that communities relied upon for grazing. And, of course, they banned fire.

European governments during the Industrial Revolution felt compelled to end rural livelihoods because self-sufficiency was a wasted opportunity for profit. Thomas Malthus, an English economist from the period, noticed that in a country such as Ireland, "where the necessary food is obtained with so little labour," the leisure time enjoyed by the public could be redirected to "create a proportionate quantity of wealth." Businessmen concurred. "When a labourer . . . lays down his spade," wrote Edmund Ashworth, an English industrialist, "he renders useless, for that period, a capital worth eighteen pence."

The effect of this new wave of legislation was to make traditional livelihoods untenable, forcing whole swaths of the population to abandon their land for jobs in cities and factories, where they would provide the cheap labor of the new economy. But there is no evidence that this was a conspiracy or a coordinated effort. Rather, the governing class of commercial elites mistook their financial incentives for a moral imperative.

By forcing people into systems of wage labor, European elites viewed themselves as improving the moral standing of the population. One English landowner expressed horror at the thought of all the hedonistic pleasures allowed by the leisure time enjoyed by rural people. The women, he speculated, "soon turn harlot, and become distressed and ignorant mothers, instead of making good and useful servants."

In the colonies, the moral imperative of fire suppression was refracted through racial hierarchies. Indigenous people who burned were "childish" or "savage," their practices "destructive." In Europe, the moral imperative reflected the hierarchies of class. Adam Smith, whose book *The Wealth of Nations* helped create the ideological scaffolding for modern capitalism, reflected, "It is remarkable that in every commercial nation the low people are exceedingly stupid." He warned that every industrial society would suffer from this class of people "unless the government takes some pains to prevent it."

Other English economists were more direct. "Races like the Celts have neither docility nor intelligence," Nassau William Senior told his students at Oxford in 1847. "They must be governed by fear."

And they were. Across Europe, as in California, the enclosure of traditional livelihoods was enforced with public executions, imprisonment, and indentured servitude. Farmers who spent their lives hunting in public woods to provide for their families and communities could now be hanged for entering those same woods. Others who incorporated fire into their practices faced prison sentences.

The era of fire suppression did not begin as a war on fire per se, but as a war against any life that could exist apart from modern capitalism. Commercial classes around the world built moral justifications for the forceful enactment of their financial interests. The result was the suppression of flames, along with the lives and livelihoods of the people who depended upon them. Fire suppression is as woven into the foundation of our modern economy as the factories, laborers, and fossil fuels that have given our society its material form.

Part II

ATTACK

CHAPTER 5

N THE PAST, HOTSHOT TRAINING WOULD LAST TWO MONTHS. HOTshots used those months to build cohesion until they could cut line as a seamless unit. But after just two weeks of training, we were called to battle an unprecedented outbreak of spring fires. A local Forest Service official bid us farewell with an ominous message: "You are entering what will undoubtedly be a record-breaking fire season with record low staffing. You will be busy. Good luck."

It was 9:00 a.m. on my first day off from the second week of training when my phone pinged with a text from Edgar. Kenzie and I were dog-sitting her parents' golden retriever, drinking coffee on the couch. I hesitated, then opened the text and read it aloud. "Come to the station. Wheels rolling in two hours."

Kenzie and I sat for a moment together in stunned silence. We had both known I would be called away, but we were just becoming accustomed to our new routine. Kenzie used our home as an office, while I left at sunrise and returned near sunset. Kenzie told me that I swore more often and chopped onions more aggressively, but otherwise it had begun to feel like a regular job.

"Shit," Kenzie said, looking at me with big eyes. "Oh, shit, shit, shit. This is real. This is really happening."

My nerves kicked in. The exertion of training had left me feverish, shaky, and shitting blood. I felt like I needed a full week of rest, and I hadn't even fought a fire yet. But the message was clear. I had two hours. I put on cotton underwear because Aoki warned that the spandex worn by athletes would melt to my skin. I pulled on my green fire-resistant cargo pants and matching hotshot T-shirt, then laced my leather boots. My bag was already packed, by the door. Kenzie put the golden retriever in the yard, loaded our own dog into our car, and was waiting for me in the driveway. Before I stepped outside, I caught a glimpse of myself in the mirror. The uniform had shifted something. I looked suddenly out of place in this pristine house. I looked like a hotshot, albeit a frightened one.

Kenzie drove me over the mountain pass and parked in the dirt lot outside the barbed-wire fence surrounding the hotshot compound. The other members of my crew were already arriving, jumping out of their trucks, laughing, giddy at the prospect of our first assignment. Scheer saw me and gave me a big smile. He beckoned me to hurry up. Our buggies were ready to go, filled with gasoline and water and military meals that would enable us to survive alone as a self-sufficient unit for two weeks in the wilderness. Kenzie squeezed my hand, then kissed me goodbye, but I still sat there in the car. It was difficult to take that final step of opening the door. This world was safe. I felt like I was about to plunge into dark water.

"I'll see you in around two weeks?" I whispered.

"Yes," she said. "I'll be here."

I still hesitated. "You'll be okay?"

"We'll be okay." She reached to the back seat to ruffle our dog's ears. "Now go. You'll do great."

※

THE FINAL DAYS OF MAY FOUND THE LOS PADRES HOTSHOTS IN PINE-top, Arizona, a small town nestled in a forest of ponderosa pines near the border of New Mexico. A plateau lifted the land out of the desert heat, creating an oasis of cool air. Hotshots are a national resource for the federal government, so unlike California's own state crews, they can be shuffled anywhere in the country to fill in gaps of personnel. At the moment, there were no wildfires in this part of Arizona, but the local crew had been called away for another assignment. We arrived to occupy the region in case impending thunderstorms brought new wildfires.

For some hotshots, this was the farthest they had traveled from home. For me, it had been some years since I spent time in rural America. On our approach, during a stop, I needed a new knife, so I popped into an outdoor retailer, where I promptly removed my COVID mask due to hostile stares. I watched a man holding an assault rifle lean over his infant's stroller, asking in a baby voice whether Daddy should buy the gun (the cashier helpfully offered that if he couldn't afford the gun, he could buy it with his credit card). To the locals, my crew must have appeared an equal spectacle. At a gas station, a woman told Barba she had seen in the news that "your people" are coming to help fight wildfires, before thanking him for Mexico's support. When we finally arrived in Pinetop, parking our buggies in the center of town, all the diners, shoppers, and bystanders paused and stared, then erupted in cheers and applause.

I found the public applause disconcerting. I hadn't fought any fires here yet, hadn't earned anyone's praise. My primary preoccupation had been to bolster my reputation among my peers. In the hotel rooms where we stayed during that seven-hundred-mile drive, I always woke early to make my roommate a cup of coffee. As a puller on Bravo Squad, my roommate was the youngest on the crew, short with a wispy mustache, and every night he video-called with girls who seemed to have put him firmly in their friend zone. He told me he had wanted to be a Los Padres Hotshot ever since he was a child. They seemed like gods. This was his second year on the crew. He was younger than many of my students back home. I served the kid coffee in bed.

This was a small gesture of humility, but it was a worthwhile way to build my alliances. With no fires to fight, everyone except the leaders woke groggy. This was because, in the hotel each night, someone would send a group message delegating a room for a party. The men congregated and sprawled out together on the beds and floor space between, shirtless and in gym shorts, to drink, smoke, and do push-ups. Márlon informed me that this was important conditioning. He prescribed me a nightly dose of six beers and several cigarettes. He was only half joking. That was the only way to mimic how I would feel once we started fighting fires every day, he said.

In Pinetop, Aoki tried to fill our time, strategizing to keep us busy, tired, and out of trouble. We began each morning with an eight-mile trail run. In my excited naivety, I had imagined we would be fighting fires the whole time, so I had only packed fire boots and flip-flops, not running shoes. I expected to be hazed for this. Instead the other sawyers demonstrated solidarity, producing their own flip-flops for the runs. Together, in our flip-flops, we managed to lead the pack. For my first time since training I felt like maybe I would be able to carve out a place in this group of men.

My relationship with Scheer and Drogo helped. While Scheer was chatty, always seemed slightly bewildered, and skimmed pictures in surf magazines, Drogo was hulking, laconic, and oscillated between obscure literary obsessions. As we drove, Drogo read Peter Matthiessen's *The Snow Leopard* while doing bicep curls with resistance bands he had attached to his seat. He and Scheer were like brothers. They bickered but were inseparable. I sat across the aisle from them in the back of the buggy, so we had more hours together than we could possibly fill with talk.

Scheer didn't enjoy silence. One day, when we were parked outside a Forest Service station near Pinetop waiting for a wildfire, he started asking me about anthropology. "So you're still a student, huh, Thomas?"

"Kind of," I told him. "Just teaching and research now. Might finish my PhD, unless I stay on the crew."

"What do you do with anthropology?"

I told him we study cultures and environments and try to figure out why people do what they do, but before I could digress into an explanation of how Edgar's governing style resembled Foucault's panopticon, Scheer stopped me. "*I know*," he snapped. "I know."

His response gave me pause. Scheer's natural disposition was that of a gentle artist, but as a third year and lead sawyer, he also felt like he needed to cultivate what hotshots call a "command presence." This made his demeanor slightly unpredictable. I was offended; I had thought our relationship was past such superficial posturing. Then I saw his face. It was sad, not stern, and I realized I had read the situation totally wrong.

It wasn't me, but anthropology that had triggered him. He told me that a woman he had loved had studied anthropology in Santa Cruz under a famous scholar named Anna Tsing (I had one of her books with me in the buggy). He had dated her for seven years, ever since high school. They became adults together. He accompanied her to the highlands of Vietnam to be the photographer for her research. But their relationship eventually fell apart. Scheer thought it was because he didn't seem to have a direction for his life. His eyes were brimming. Drogo turned around in his seat to watch Scheer. He looked like he might cry as well.

Scheer tried to compose himself, clearing his voice. "But that's life, right? I'm a hotshot now. I'm finally the person she always wanted me to be." He turned to stare out the window. "I just became that person too late."

Whenever people outside the crew asked me why anyone would work a job like this, where the risk is so high and the pay so low, I never knew quite how to respond. During training, a retired colonel from the U.S. Special Forces had visited our station to give us a speech, during which he informed us that, like him, we all must have a genetic predilection for adrenaline and risk. I felt this was nonsense. Matthew Desmond, a Princeton sociologist and former wildland firefighter, found that firefighters actually prefer working with colleagues who are risk averse and capable of calm, rational thinking, which keeps everyone safe. For Desmond, decisions to enter careers in wildfire aren't just about individual choice, but about social structures of inequality that shape the choices available to us. These choices are funneled by economic bleakness, including "the land loss taking place in rural America, the inability of small-town communities to carry out economic development strategies in the face of global demands, and the rising costs of higher education." All else being equal, with wildland firefighters possessing at least as much work ethic and intellect as the wealthiest individuals in society, it is the calcification of social class that determines which "bodies, deemed precious, are protected," and which, "deemed expendable, protect."

Personally, while I felt like it was lazy for the Special Forces officer to reduce our motivations to genetics, I also felt it was incomplete to limit them to class. When I asked Aoki why he became a wildland firefighter, he told me he was on his couch one day, eating potato chips and watching television, when the news panned to a wildfire. He was in his early twenties, thought the job looked cool, and found that he was very, very good at hiking, so he stuck with it and became the best. For Red, this was the only job that allowed him to spend most days outdoors with a community akin to family. Drogo said he enjoyed living like an outlaw. For Barba and Márlon, as immigrants, this was a job that offered them a foothold in American society, but it was the favorite job either of them had ever worked. "As long as you're in shape," Márlon told me, "this job is a helluva lotta fun."

My own motivations were just as cryptic, even to me. Three years before, when I had arrived in California, I didn't have a job, and I didn't have very much money. My research experience in fire was the only conceivable bridge between my past life and my new home. But my own economic desperation floated atop a sea of other motivations: the allure of physical challenge, the joys of camaraderie, the fulfillment of social respect, and the intoxicating sense of wonder I gain when learning a new trade. While we all had rote explanations we could deliver neatly packaged when pressed, I often felt that these narratives functioned less as hard truths than as mechanisms for granting a semblance of order to the otherwise chaotic matrix of personal histories and experiences that sweep us all through the labyrinth of life. Somehow, these paths had landed twenty of us here, in a pair of hotshot buggies in a forest outside Pinetop. Inequality and social class had certainly played a role, but so had love and broken hearts.

SOMETHING RAPPED AGAINST THE WINDOW OF THE ALPHA BUGGY, JOLT-ing me from my thoughts. Due to Scheer's fragile state, we ignored the sound until it came again. The buggy occupied by Bravo Squad was parked parallel to us. Through the window, the hotshots were jeering, giving us the middle finger. Barba's face suddenly appeared, and he launched a carrot at us, which missed again, knocking against the buggy. This brought Scheer back to his normal self. He handed me his own carrot from his sack lunch. "Give it back to them, Thomas."

Without thinking, I chucked the carrot out the window, right when Barba's face appeared again. The carrot hit him square in the forehead. A shadow crossed Barba's face, which spread to the others on Bravo Squad. Slowly, ominously, they raised their windows, disappearing from view.

I turned back to the others in my buggy, triumphant, but found them staring at me as if I had volunteered for my own execution.

Drogo shook his head. "Big mistake, Thomas."

"I can't believe you did that," Scheer said. "The guy's crazy."

"You told me to!"

"That doesn't mean you do it!"

Márlon turned around from up front in the buggy, where he had been watching a movie on his phone. "Don't worry, homie. We got your back."

Unbeknownst to me, Barba enjoyed immunity from the pranks of the crew. His time serving in the Marine Corps had taught him that exponential and limitless escalation was the only way to stay ahead of a conflict. During his first year on the hotshots, he stuck to this code. Legend had it that, after he learned someone had tied a phallic sausage to his pack where it dangled all day while he worked, he put bad cheese in that person's sleeping bag. As the retaliations continued, he eventually snuck into the hotshot compound in the middle of the night with a torque wrench, intent on removing the tires from the personal trucks of his opponents. Now, because of me, but also because we were bored, the two hotshot squads were at war.

Barba and his allies were patient. They waited for several days, allowing the trail runs and forest maintenance work to seemingly ease tensions, before hosting a movie on a television they had rigged up in their buggy. The idea of killing time in front of a screen was tantalizing, and they knew it, so they sent my hotel roommate, the young guy with the wispy mustache, to invite me to join. Because I served him coffee in bed, I assumed I could trust him, but when I pulled myself into the back of Bravo Squad's buggy, I was temporarily blinded by the sudden move out of daylight. A dozen hands reached from the darkness, gripping my shirt to pull me inside.

They were trying to "bump" me. This is a hotshot tradition in which if a member of one squad can be lured into the others' buggy, all eight members grab him, pull him inside, lift him over their heads, and slam him repeatedly against the ceiling. I never understood how hotshots accomplished this without breaking the victim's nose. Maybe they did break it. I didn't want to find out. I twisted and pushed and thrashed free of their hands, throwing myself out the door to sprawl in the dirt.

An attack on me was an attack on my squad. Márlon, true to his word, offered support. He was a solid ally. Márlon had been a firefighter for eight years, with five on this hotshot crew, and he was unshakable. Throughout the season, as I got to know him better, he exhibited the same chill demeanor when we were nearly trapped in burning trees as he did on quieter days, when he would prop his phone against a shovel so we could watch movies together. The others on the crew speculated that his tolerance for danger came from his childhood in Nicaragua, where he told me he once outran the police on a motorcycle, with his girlfriend on the back, by weaving through a herd of cows and finding a picturesque hideaway on the shores of an emerald lake. Márlon exhibited a natural grace that belonged more in a tuxedo than a firefighter uniform. But he was also a scoundrel. Once, according to hotshot lore, he saved Barba from a bar fight by using his belt to whip away three assailants. Nevertheless, his loyalty to his squad came first. After he heard that Barba had tried to bump me, he meticulously gathered hundreds of biting insects and set them loose in Bravo Squad's buggy.

As the prank war escalated, Aoki pretended not to notice. After our runs, he tried educating us, taking us into the forest to teach us how to cut down dead trees, which had the added benefit of helping with the jurisdiction's forestry projects. But as the days dragged by, Aoki needed to be increasingly creative. He assembled a miniature mock village out of sticks and tools and tape, then called us over, one by one, with radios, so we could gain experience acting as incident commanders of an imaginary newly ignited fire. I learned that wildfire names, unlike hurricanes, don't follow the letters of the alphabet, but are titled by the first responders on scene. While typically firefighters name them for nearby landmarks, the implicit creative license also accounts for some of the more poetic monikers: the Midnight Fire, the Bogus Fire, and the Sour Biscuit complex. (An incident goes from being a "wildfire" to a "complex" when multiple fires are burning in the same vicinity and fall under the same operational command.)

So, amid the prank war, Aoki tried to keep us studious. When he ran out of lessons, he set us to sharpening tools and tending to our chainsaws. But each day without a fire further darkened his mood. When my chainsaw was entirely disassembled, with nuts and bolts and tools spread along the side panel of the buggy, Aoki appeared from nowhere, peering over my shoulder.

"What are you doing?" he asked in a sharp voice.

"Cleaning my chainsaw."

"If we got called for an initial attack, would you be ready to go in thirty seconds?"

I cowered. Aoki was the one person on the crew we all wanted to impress. The answer to his question was obviously *no* but, by now, a call to fight a fire seemed as farfetched as Edgar's text to mobilize had seemed a week before in Santa Barbara. Hotshots like to say that 90 percent of the job is boring, but the other 10 percent makes up for it. Drogo compared our situation to his being stationed in Hawaii as an army infantryman during the war in Afghanistan. "When you're trained for something, you don't just want to practice doing it. You want to do it."

And so we waited, hoping for an initial attack. An initial attack, or IA, is the zenith of fire suppression operations, allowing us to be the first crew on the fire's edge. "That's what hotshots live for," Scheer told me.

Then, just when an initial attack seemed a distant dream, when the routine of running and practicing and pranking had softened my nerves, and when it seemed inevitable that we would sulk home as faux heroes—just then, we heard a noise. It started in a high pitch before dropping in frequency, zapping us all like an electric shock coming from the radio in Aoki's truck. A voice followed the sound, announcing a lightning fire in the Apache-Sitgreaves National Forests. Within thirty seconds, we were gone.

THE AMERICAN WEST IS FULL OF PYROPHILES, OR FIRE LOVERS—SPECIES of plants, animals, and fungi whose existence depends upon their ability to follow ignitions. Of these species, the fire beetle is perhaps the most tenacious. These beetles are black, the size of a fingernail, and are equipped with heat receptors the width of a human hair. Their receptors hold liquid that expands when absorbing radiant heat, allowing the beetles to detect flames from over one hundred miles away. Wildfires act like magnets, pulling the beetles in swarms of millions, where they mate amid the flames, waxy bellies dispelling heat as they bore into charred wood to lay their eggs. In California in the 1940s, football games were occasionally disrupted when the collective embers of spectators' cigarettes attracted beetles that, finding no trees, attempted to burrow into human flesh. Axel liked to joke that Aoki was a fire beetle in human form. Sometimes when Aoki was making phone calls with higher-ups, trying to find assignments for us, Axel would point his fingers over his helmet like antennae and prance around, emitting a high-pitched beetle whine that ended in our supe's name, "*Aaaoookiiii.*"

Now, on the first day of June, the joke seemed apt. Aoki led our green buggies in his white truck along a rural mountain road, toward a draft of smoke that blended with clouds in the shifting light of late afternoon. Inside the buggies we sat in silence, engaged in our own obsessive rituals of preparation: hydrating, checking our gear, making sure our packs were tight, our chaps were unrolled, our yellow fire-resistant shirts were ready, and our gloves were accessible. I closed my eyes and completed a round of circular breathing to calm my nerves. When I finished, we were only halfway there. The drive to the fire took two hours. The time seemed to stretch forever. I didn't understand how a fire burning for two hours without intervention could possibly be contained.

Like the fire followers of the natural world, hotshots are drawn to the annual cycles of combustion in a giant counterclockwise motion around the American West. The first ignitions typically occur in late spring and early summer, when monsoons sweep Nevada, Arizona, and Utah with lightning. Later in the summer, when high elevation snowpacks begin to melt, the fires spread to the Rocky Mountains of Colorado, Montana, and Idaho before moving down the coast from Oregon to Northern California. The end of the fire cycle would land us right back home in Santa Barbara when, in autumn, the end of the dry season coincides with strong winds that blow wildfires down the mountains toward our homes. But among the fire followers of the American West, we were unique, because we followed the cycle of fire in order to break it.

We found the base of the fire at a bend in the highway. We parked on the shoulder along the edge of a sheer cliff. The other side of the road rose steeply into the forest. To our disappointment, several crews had arrived before us. A Forest Service engine crew was watching for traffic, but the region was so remote that there was none. A tribal Apache crew was already up one flank of the fire. We could hear them cutting line. As soon as the buggies stopped moving, we swept out the back door, pulled on our packs, hefted our chainsaws and hand tools, and formed a single-file line. We stood silently, which was part of our tacitly agreed-upon performance in the presence of other crews. Aoki disappeared into the trees to scout. Edgar stood at the head of our line, briefly appraised us, then led us up the slope and toward the fire.

When we approached the fire, the scene was far from what I had imagined. Instead of a wave of flames demolishing everything in its path, the fire we found crawled gently, low to the ground. I felt like we had stumbled upon a living creature, a local herbivore wandering curiously about, searching for forage. There was no accelerating head fire, only a series of tendrils that crackled and coiled through a mosaic of boulders and drainages. In some places, it ran out of fuel and crawled off in new directions, while in others it simply went out. Within the parts that had already burned, the old trees were mostly alive, and the ground around them was a soft carpet of white ash. I felt like I was observing fire as it once was.

We split our squads, with Bravo attacking from the bottom flank and Alpha hiking to the top so we could cut a path through the vegetation and scrape a line of dirt, moving toward each other. There was little danger and even less heat; our task was merely to follow the smoldering edge to remove the sparse vegetation and scrape away the pine needles. Alongside where we walked, the flames ate their way through layers of leaves and logs, accelerating decomposition without destroying essential nutrients in the soil. Occasionally, toward its center, the fire crawled into the canopy and consumed a tree in a burst of light and heat. Across the American West, these dynamics are essential for ecological cycles of life. Gentle flames crack the cones of many conifer species, dispersing seeds, while periodic intensity creates gaps in the canopy for saplings to grow. In California, for each of the state's seven thousand species of plants, there are multiple species of fungi. In shades of orange and red, mushrooms unfold throughout ashen forests. Some, like the *Pyronema*, feast on charcoal, transforming it into nutrients for the next generation of microbes that begin the process of ecosystem recovery. In other regions, after a fire, hundreds of species of herbaceous plants have a moment in the sun: poppies and pussypaws, ferns and clovers, lotus and lupine and Linanthus. These dazzle slopes in color, sweeping mountains with aromas that attract honeybees and butterflies and hummingbirds until the whole land vibrates in a pollinating hum. The seeds of many species only open when smoke triggers germination.

In retrospect, the idea of attacking such a benign forest creature feels malevolent, almost contemptible, like shooting rabbits for target practice. But there was a copper mine several miles away whose operations would be paused if this fire was allowed to spread. Besides, I wasn't thinking about any of this as we worked. I was focused entirely on my own performance, hoping to maintain whatever status I had eked out of my first few weeks on the crew. Nevertheless, I was sloppy. I kept misreading the edge of the fire, either cutting too close to the embers and sending branches into them, or moving too far away where my cuts were worthless. Scheer, between his own cuts, stopped to glance back at me, frustrated, making indecipherable chopping motions with his gloved hand. But it helped that the rest of the crew was sloppy too. Another first-year hotshot couldn't handle the smoke—he puked, cast his tool to the ground, and ran away, sitting on a stump to gasp in panicked breaths. Such was the pressure of this social universe that, rather than empathizing with him, I found myself grateful that the crew's ire turned toward him instead of me.

Within the hour, we tied in with Bravo Squad and began "cleaning" the line, which now looked like a hiking trail, by carefully removing any leaves, twigs, or other organic matter that could allow the fire to creep across the dirt. Then we patrolled inside the burned area, searching for heavy logs, saplings, or bushes that could cast sparks across our line if the wind picked up. Edgar instructed us to toss these into piles, called "bone piles," deep within the burned area where they posed no danger. Soon, our bone piles burned like bonfires within the waning light of the forest. When I stopped to enjoy the scene, Aoki appeared from nowhere, chastising me for losing vigilance. The work wasn't finished until we departed.

Normally, we would spend several days "mopping up," monitoring the fire until it was "dead out," which meant no heat could be detected within the confines of our line. But fires in the region were intensifying. The next day, we received a call to the San Carlos Apache Reservation. We followed blacktop freeways down from the mountains, through red rock canyons into stifling heat, where dirt lanes narrowed like capillaries until we reached the end of the road on a dusty hilltop. Two different fires smudged the horizon. A helicopter landed to shuttle us and our weapons into the reservation. With mounting foreboding, I could not shake the sense that we were invading Indigenous land with weapons of ecological war: night-vision goggles, chainsaws, incendiary guns, grenades, and flame torches. For the next twenty-four hours, as we suppressed our second fire, retardant-bearing airplanes roared overhead, and our chainsaws screamed in the brush.

The final night of our first assignment, as I tried to sleep under a sky blue with starlight, my chainsaw rang in my ears like a ghostly, violent echo. My hotshot crew, I realized, was not just breaking ecological cycles, nor were we simply mimicking the practices brought to this land by colonizers; we were direct inheritors of their legacy of fire suppression. We were specialists in the practices and mentalities that led America's environmental relationships awry. By making an enemy of fire, we'd made an enemy of the land.

From that time on, the fires would grow, and we would attack them relentlessly.

CHAPTER 6

A THUNDER OF HOOVES AND CRACK OF RIFLES ANNOUNCED FIRE suppression in California under American rule. The year was 1846, and the spring was young. The rivers of the north were swollen with snowmelt. Just twenty-five years before, Mexico had gained its independence from Spain, and with it, control of California. As the Mexican-American War seemed increasingly imminent, Army captain John C. Frémont rendezvoused in Kansas City with Kit Carson—a frontiersman already infamous for his barbarity—and rode west to the coast. Unabashedly ambitious, Frémont wanted to be there when war was declared so he could be the one to seize California for the United States. Now, on the morning of April 5, 1846, Frémont and his men surrounded a gathering of Wintu people. Frémont gave the order to attack.

At the time, even after the brutality of the Spanish colonial system, approximately 150,000 Indigenous people survived in California. Though the Spanish had effectively stamped out the flames of the coastal south, in much of the state the lives of Indigenous people were comparatively unaffected. Smoke still curled over hills and forests, with cultural fires creating habitats for elk to roam the plains. Flocks of geese still darkened the skies, and shoals of salmon still swam in the rivers. In the year of 1846, because of the persistence of California's Indigenous people, most of the land's ecosystems remained in balance with fire. But with the arrival of Captain Frémont and his men, this would soon change.

Frémont's contingent had hunted this group of Wintu people on the recommendation of a nearby rancher, who reported that they were gathering in great numbers. The rancher described the gathering as a war party, which elided the fact that most of the people present were women and children. More likely, the Wintu had congregated to fish for salmon and to celebrate the end of winter.

Yet determined to win favor with the American settlers of California, Frémont and Carson were not concerned with investigating the purpose of the Indigenous gathering. "Found them to be in great force, as was stated by those who had requested Frémont's help," Kit Carson recalled. About one thousand Wintu people were gathered at a bend in the Sacramento River, surrounded by water on three sides. The Americans sealed off their escape route. Dressed in buckskin, armed with rifles and butcher knives, Frémont's men opened fire.

The Americans released their first volley from a distance. "The order was given to ask no quarter and to give none," recalled one soldier who was present. Out of range of Wintu weaponry, the attackers fired until their guns clogged, then charged into their midst. "And then commenced a scene of slaughter which is unequalled in the West," the soldier continued. The Wintu "were shot down like sheep and those men never stopped as long as they could find one alive."

The Wintu tried to flee. Most of Frémont's men lined up on the riverbanks to shoot those attempting to swim across the frigid water. Kit Carson pursued those running toward the foothills.

That day, Frémont and his men murdered as many as one thousand men, women, and children. "It was a perfect butchery," Carson concluded. He considered the operation a resounding success. After the massacre, the men continued their journey north, then south again, killing any Indigenous people they encountered along the way. They were following orders from Frémont to "shoot Indians on sight." This massacre was a prelude of Indigenous life in California under American rule, and the coming genocide that would almost entirely erase healthy fires from California's land.

Four years after the massacre, when California gained statehood in the aftermath of the Mexican-American War, John Frémont leveraged his military reputation to get elected to one of California's first Senate seats. In that brief interim, settlers had discovered gold in the Sierra Nevada mountains, and the white population had tripled, outnumbering California's Indigenous people for the first time. Many settlers arrived heavily armed.

Frémont's predilection for violence and dehumanization were not unique in this first government of California. In the first legislative session, the governor said that a "war of extermination" would be waged "until the Indian race becomes extinct." This intersected with a bill that criminalized the use of fire by Indigenous people. The pairing of genocide and fire suppression was no coincidence. According to Bill Tripp, an Indigenous fire specialist and member of the Karuk tribe, colonizers blocked cultural burning practices as a deliberate tactic of ethnic eradication. "It became part of the policy to remove that connection to the food systems," Tripp said. "When you're going through a cycle of genocide and people are trying to remove the Indigenous component from a place, Indigenous burning practices become a logical target."

This method of destroying Indigenous economies fit into a pattern of U.S. genocide against Indigenous people. In the Great Plains, for instance, where buffalo long outnumbered the U.S. population and provided the backbone of Indigenous livelihoods, a general advised Congress to mint a medal that sported a dead buffalo on one side and a dead Indian on the other. Within the decade, white soldiers and vigilantes—supplied with government equipment and weaponry—had nearly driven buffalo extinct. By removing people's connection to their food systems, just as the Spanish had along California's coast, America was opening the land to seizure and exploitation. In California, fire had always connected people to their food, and Americans set about its suppression with unprecedented brutality.

The burn prohibition in California placed Indigenous people in a particularly difficult bind. They could either continue to burn and be prosecuted as criminals, or they could abandon their traditional economies, risk starvation, and be prosecuted as vagrants. The penalty for either, as outlined by the same piece of legislation, was slavery.

This was a boon for settlers seeking to profit from California's natural wealth. The government kept the legal justifications for taking slaves vague, and the state's criminal code clearly prohibited Indigenous people from defending themselves against any white person in court. This allowed for white settlers to enter Indigenous villages, kill most men, and kidnap women and children for labor. The justifications were always flimsy—a lost cow or misplaced horse blamed on "Indian theft"—and the murders and kidnappings were rarely subject to judicial scrutiny. In the first years of California's statehood, historians estimate that settlers enslaved some twenty thousand Indigenous people. Most massacres surrounding these acts of enslavement were not documented.

America's campaign of extermination was devastating for California's Indigenous people. By criminalizing fire, crippling traditional economies, and enslaving many survivors, the settlers forced themselves into parts of the state that had remained mostly intact throughout the past century of colonial rule. Yet if the violence had stopped there, Indigenous people could feasibly have maintained a balance between fire and California's ecosystems. The legislative ban on cultural burning, however, laid the groundwork for a much larger project of wholescale genocide, which would become the most significant act of fire suppression in American history.

According to historian Benjamin Madley, the escalation of the genocide was triggered when a group of men of the Pomo tribes of Clear Lake revolted against and killed the two white men who had forced them into slavery. It would have been difficult to imagine the maelstrom of violence this small act of justice would unleash. When news of the killing reached Sonoma, some seventy miles south, it sparked more than fury among the settlers. It sparked fear—the fear of insurrection. The settlers responded to this fear. Led by Moses Carson, the brother of Kit, the U.S. 1st Dragoons saddled their horses and galloped toward Clear Lake.

When the Dragoons rode north, they did not bother to distinguish one tribe from another. En route to exact revenge for the killing of two white men, they razed two villages and killed all the inhabitants.

Upon arrival at Clear Lake, the Dragoons noticed that a large group of Pomo people had sought refuge on a small island. One officer observed that it was "a perfect slaughtering pen." Carson sent an envoy to the military, requesting boats. When the boats arrived, the Dragoons attacked. According to the *Daily Alta California*, a regional newspaper, "They . . . poured in a destructive fire indiscriminately upon men, women, and children. . . . It was the order of extermination fearfully obeyed."

With the fears of the settlers stoked, precedents for massacres set, and legal impunity more or less granted, the public furor for genocide accelerated quickly. In 1853, a newspaper in Yreka, *The Mountain Herald*, proclaimed, "Extermination is no longer even a question of time—the time has already arrived, the work has been commenced, and let the first white man who says treaty or peace be regarded as a traitor and a coward."

Many local governments took this new phase of killing into their own hands. In Humboldt County, the citizens of Uniontown and Eureka voted for a tax to be levied on residents "to prosecute the Indian war to extermination." Sometimes the state and federal government outsourced the labor, reimbursing the costs that private militias accrued in the genocide. One county paid fifty cents for each Indigenous person's scalp and five dollars for every head. In a single trip, a man delivered twelve. In 1868, the editors of the *Lassen Sage Brush* proposed a five-hundred-dollar reward for "every Indian killed."

Taxpayers were more than happy to foot the bill. According to historian Brendan Lindsay, by 1854, after just four years of statehood, California had already spent $924,259 reimbursing militias for Indigenous genocide—$32 million in today's currency. Other historians place the figure higher. The expenses were so vast, the massacres so prolific, that they threatened to bankrupt the young state. Legislators resorted to issuing war bonds to pay for the killings.

The full scale of the genocide will never be known because many of California's Indigenous groups cremated their dead. Yet population statistics from the era are telling. In 1769, when Junípero Serra first walked north with Spanish conquistadors to establish the California missions, the Indigenous population of the region numbered around 300,000. By 1850, when California gained statehood, that population had decreased by half, to 150,000. Just forty years later, in 1890, the population of California's Indigenous people had fallen to 16,624. In a mere forty years, the citizens of the United States killed 90 percent of California's Indigenous people. Historians Robert Hine and John Mack Faragher concluded that California represents "the clearest case of genocide in the history of the American frontier." If California is the clearest case of genocide in U.S. history, then genocide is the clearest case of fire suppression. In 1850, when the genocide began, most of California's land was still touched by Indigenous flames. By 1890, those flames were mostly gone, extinguished along with the lives of the people who lit them.

❧

NOT YET TWO MONTHS INTO THE FIRE SEASON, MY HOTSHOT CREW HAD fought a desert fire in Nevada, stopped an Arizona lightning burn, helicoptered into Big Sur, and endured the hottest local temperatures ever recorded outside Palm Springs. After another week extinguishing a fire near the border of Oregon, we were dirty, smelly, and tired. Our skin was pocked with bites from fire beetles. My heels were raw with blisters. Scheer's feet were cracked with fungus. Heat illness sapped our strength and clouded our minds. But there was no time to rest. On the afternoon of July 7, we drove south into the Sierras to fight our first megafire of the year.

On our approach, the Sugar Fire prowled in the mountains like a bear woken early from hibernation, scavenging hungrily, growing quickly. We pressed our faces to our buggy windows, landscape zipping by, several miles of pasture between us and the smoke. Márlon was the only one who seemed uninterested, but even he removed his headphones when he saw the plume bending over the peaks. "Man, you normally wouldn't see fire act like this until September," he noted. "Gonna be a crazy year."

Márlon's observation echoed what scientists were saying across the country. Daniel Swain, a climate scientist at UCLA, had reported two days prior that "we are seeing June fires behave like August and

September fires. And we're seeing August and September fires behave in ways we didn't really see at all historically, except under the most dire conditions."

As our hotshot buggies rumbled closer to the plume, sixty-five thousand firefighters were actively battling forty-four blazes that had already burned seven hundred thousand acres across the American West—more than double the historical average. Much of this extreme fire activity was driven by recent heat waves that had brought regional temperatures over 11 degrees Fahrenheit above all-time highs. Scientists were cautioning that the neat graphs representing a steady uptick of carbon in the atmosphere failed to capture these nonlinear changes happening on the ground. Wildland firefighters face more hazards than most, the scientists warned, because their survival depends on knowledge that is calibrated to a world that no longer exists.

In the buggy, the Los Padres Hotshots appeared more excited than concerned. For the past several days, on the border of Oregon, my crew had gone stir crazy after Axel contained our wildfire by leading a bulldozer for two miles along the flame front until it was fully contained. The victory had been followed by several days of monitoring the fire to make sure it didn't cross Axel's bulldozer line. The lazy days of watching wind spin ash into dust devils seemed to take a greater toll on my crew's well-being than the prior weeks of sweating and sleeping in the dirt. One afternoon, I had found Drogo sitting against a tree stump, chain-smoking cigarettes and finishing his third root beer. "This is a reflection of my mental health," he told me, offering a drag. Returning to the vehicles, I found Scheer musing that, in the hours we had been sitting in the buggy, he could have flown all the way to Bali to surf. "The airplane would have air-conditioning and booze," Scheer mused, staring at the seat in front of him. "And it wouldn't smell like farts." The hotshots were happiest when fighting fire, and the Sugar Fire looked promising.

On our approach, we rumbled into Portola, a mountain town bisected by vacation mansions and local shacks. A shirtless man with neck tattoos sat on a porch decorated with Trump signs. He offered a fist salute as we passed. We parked at the incident headquarters, a school repurposed for suppression operations. The fire had only ignited several days before, but the headquarters was already bustling.

From the buggy window, I watched Aoki stroll through the headquarters. His presence sent ripples through the crowd, with workers pausing to watch him pass and forestry officials converging at his side. This was normal, Scheer told me. "It's always like, 'What do we do? What do we do? Help us, Aoki!'" By the time Aoki was ushered into the classroom-cum-command center, he was surrounded by a flock of personnel he seemed to have known for years.

While Aoki was getting briefed, we had about thirty minutes to brood on our future. Drogo thought we would take helicopters into the mountains. Scheer predicted night operations. Márlon had been around long enough that he didn't bother speculating. "I don't fuck with the crystal ball," he said. It was impossible to predict where Aoki would send us. The only way to stay sane was to be prepared for anything.

Even after this advice, I wasn't prepared for Red's shout when Aoki returned to his truck. "Time to buck up, fellas!" Red called from the driver's seat, firing up the diesel engine. His face was flushed with excitement. "We're goin' in!" Everyone cheered. Edgar cranked Metallica.

Leaving town, we followed a grid of dirt roads along the base of the mountains until we were adjacent to the fire column. Aoki pulled over, stepped out of his truck, and spread a map along the hood to study the terrain. Golden fields stretched about a mile to the slopes, which climbed in boulders and brush until forests darkened the higher climes. He shook his head, glancing between the mountains and the map. The smoke stacked over it all like an atomic detonation.

A radio crackled, ordering all crews to retreat from the fire's edge. The wind was shifting, the fuels were too dry, and the fire was becoming unstoppable. The most we could do was protect the homes ahead of its advance. Aoki folded his map, hopped into his truck, and we caravanned into the mountains. Like that, our plans changed. We weren't going direct, but we were going as close as possible to the fire's head so Aoki could devise a new strategy.

Through canyons filled with sycamores and ferns, up ravines and into the heady scent of pine, the road brought us deeper into the mountains. Cresting a ridge, we gained a bird's-eye view of the fire. The head was bulbous and shifting. It seemed alive, slithering down the valley. As it consumed the forest, the core of the blaze pulled oxygen toward its own center, forcing the surrounding pines to whistle and bow until they popped like matches and joined the growing inferno. Airplanes had painted the ridges pink with fire retardant, but the fire was burning so hot that it rolled effortlessly through. Helicopters flitted around the smoke column, appearing as small as mosquitoes.

We rolled over the ridge, back into the canopy, caravanning through the shadows until we emerged in an alpine watershed. We approached a cluster of homes. A dust cloud came from the other direction, residents' jeeps and trucks fleeing the fire. One man hung out his window, screaming "LP!" and waving his thumb and pinkie finger in a shaka sign. The hotshots were already cranked with energy. Being recognized by an evacuee put us over the edge. As we neared the first house in the meadow, Scheer dove from the buggy before we parked. Bravo Squad was already at work, led by Axel, unplugging propane tanks and cutting trees and bushes from around the homes.

The fire was above us now, boiling over the ridge with wings of orange smoke that coiled into the sky. The largest trees catapulted firebrands from the mountaintop, causing new ignitions that merged and pulled the flames toward us in jolts. Historically, this meadow would have been spongy with snow runoff. It would have been unignitable, serving as a natural check on the fire's advance. This year, however, California had received just over half its average snowpack, so the meadow was dry.

Before I could even pull my chainsaw from the side panel, I was stopped by Red's voice.

"Load up!" he yelled. The fire had established itself in the meadow and was sprinting through the watershed, straight toward us. Aoki was already in his truck.

We retreated back into the forest. In the darkness, I could not shake the sensation that we were hiding within our own coffin. Though it was only late afternoon, the world had gained a reddish hue under a persistent rain of ash. The fire was near. I could hear it, rumbling in the mountains. It was only a matter of time before it crossed the meadow and entered these trees. The trees rustled, pulled by the fire's breath. We parked the buggies near another hotshot crew and milled around while Aoki scouted a new course of attack.

I was nervous, but the rest of my crew seemed to be having fun. Their eyes were bright, speech staccato, movements emanating a kind of physical energy that, in another context, might have suggested stimulants. We were wired by adrenaline, and it made us unnaturally animated.

"Dude, do you fart in your sleep?" Barba asked Jack.

"Probably. Why?"

"I woke up last night and was choking on this horrible air."

"It was probably me."

Jack yanked on his beard, as he always did when excited. He speculated that we'd get shipped to Canada to help with the wildfires in British Columbia. They were having some of their worst fires ever too. Tyler, the stocky mountain man who was our lead sawyer, stopped polishing his saw long enough to laugh at Jack. "Pending felonies, don't like to drive sober, don't trust vaccines, and to *hell* with proof of citizenship—send the hotshots to Canada!"

Axel, sitting in the driver's seat of Bravo buggy, looked up from the 1920s travel memoir he was reading to join the joke against Jack. "*Passport?* Why in the doggone hell would I ever need a *passport?* 'Murca is the best doggone country in the world and I ain't never leavin'."

Tyler exaggerated Jack's country accent. "Darn Canadians, with their maple leaves and flannel shirts. Speakin' that funny English."

Jack laughed with the rest of us. It was almost enough to keep me from noticing that smoke was filling the understory of the trees.

Then a radio cut through our noise. The forest regained its eerie quiet, a whisper of trees, a mumble of combustion.

The static became a voice. Someone was scouting the fire near Aoki. "Yeah," the voice said. "I think this fire's just gonna do what it wants today."

BOOM. A home exploded nearby. Drogo elbowed Barba. "How's that PTSD holding up, Marine?" Barba grinned but didn't respond. We were all trying to hear Aoki's response to the voice on the radio as the forest shifted to a ruby glow.

"Roger that," came Aoki's voice through the static. "It's looking like it's gonna be a Hail Mary burn op."

His words sparked movement, as if releasing our tension. The only thing worse than standing in the path of a megafire was doing so without a plan. Everyone hurried to the side panels of the buggies to distribute shovels and pickaxes and prepare chainsaws. I was unsure of what to do.

"Ready, Thomas?" Barba appeared beside me, tool in hand.

My expression must have been blank. I had never been a part of a burn operation before. To fight fire, we removed fuel. I knew that. Normally, to remove fuel from the fire, we cut line, using chainsaws to remove trees and brush from the edge of the flames before scraping the ground down to soil. Fire can't spread through soil. But megafires are so hot and volatile that these traditional techniques are too dangerous. Even if we could survive along the fire's edge, the flames would blow over our line.

Barba explained to me that burn operations are another way to remove fuel ahead of the megafire's advance. Instead of cutting fuel away from their edges, hotshots light fires around them at a distance. This technique works because megafires devour so much oxygen that they create an atmospheric vacuum. That vacuum would pull our flames toward the megafire, devouring all the fuel in its path. If we lit our fire along a fuel break, such as a road, only the forest between the road and the megafire would burn. The head of the megafire, with all the surrounding forest burned, would stop spreading. It would starve. That, at least, was our hope.

Many firefighters think of this as fighting fire with fire or, in the words of the first chief of the U.S. Forest Service, Gifford Pinchot, using the tactics of the enemy against them. But I was reminded of the research of Omer Stewart, that old anthropologist who documented the traditional burn practices of Indigenous people in California. When people in the Pomo tribes would set fire to the land, they would use this same technique to contain escaped burns. They would run to a ridge above the escaped burn, light the ridge on fire, and guide the controlled fire downhill toward the wild one. But they described fire as a reciprocal part of an environment they were invested in maintaining. I wondered, now, how this same technique would work when translated through a military mentality of attack.

Edgar marched by, interrupting Barba's impromptu lesson. "You're on a tool tonight, Thomas. No need for chainsaws." Edgar's eyes were sharp, his movements precise. I didn't reply, because I saw that he was calculating conditions I could neither see nor comprehend as he prepared to set the trees around us ablaze. "Line it out!" he called. "Twenty-foot spacing!"

Barba caught my eye and pointed for me to stand behind him. We assembled single file, then walked into the forest directly toward the megafire. I could not see the fire yet, but I could hear it like a distant waterfall.

"You know what we're doing, Thomas?" Barba kept his voice low so only I could hear.

"Help me out."

"See that road down there?" I looked to my right and saw a dirt road about fifty feet away. "That's our fuel break."

"Gotcha." The road was ten feet wide. Instead of cutting our own fireline, we could use the road for the same purpose. The road led to a meadow. The head of the wildfire was on the other side of that meadow. If we could carry a controlled burn along the road and into the meadow, we would create a giant fuel break shaped like the number seven. This would, at the very least, steer the fire away from the homes in the forest.

"The Plumas Hotshots are gonna burn off the side of the road," Barba said. "We're gonna be holding it."

On the other side of the road, I could see dark figures striding through the forest, wielding torches that sent flickers through the dusk.

"How're we gonna hold it?" I stopped walking. Barba stopped twenty paces ahead of me. Our entire crew was spread in a line that stretched almost a half kilometer through the woods.

"Eyes in the green, homie."

"Gotcha. Eyes in the green." I understood what that meant. To keep our controlled fire between the road and the megafire, we stood with our backs to it. This allowed us to scan the forest floor ahead for embers that could float across the road, from the controlled fire, into "the green." For hotshots, the world is split into two categories: *the green*, which is anything that hasn't burned, and *the black*, which has already burned, making it unignitable and, by extension, relatively safe. The green is where no hotshot wants to be. If a stray ember caught in the green, it would become a spot fire and could trap us in the burning forest. So: eyes to the green, and don't look back.

But it is a difficult thing, I learned, standing alongside a forest fire with eyes averted. I sensed the hiss of flames that *whoomf*ed into the canopy behind me, the heat against the nape of my neck. Turning away felt like gazing into the face of a wolf, then closing your eyes. So, I snuck peeks. Behind me, a woman from the Plumas Hotshots walked through the burning forest between the road and the fire. Her hair was dreaded under a blue helmet, eyes white in a face of soot. She dripped flames from the end of her torch like water from a garden spout.

A grunt jerked my attention. Barba had caught me watching. He pointed to his eyes and toward the darkening forest. *Eyes in the green.* His face was stern. In that moment, Barba could have made a show of scolding me in front of the others, an assertion of hierarchy. Instead, he winked. I saw that he was sneaking peeks too, filming the flames on his phone without looking at them.

"Moving!" shouted Tyler from up the line. "Fifty-foot spacing!" Barba echoed the message, and I did the same, words passed from one to the next until they faded through the trees. I waited until Barba moved, then I followed him, letting the space between us stretch farther. As the head of our controlled burn moved, we followed it. The fire front was leaping two hundred feet above the canopy, but the back was quiet now, all embers and smoke. We had already wrapped a half mile of ash around the head of the megafire. If we kept this up, there would be nothing left for the megafire to burn when it arrived.

I was the third from the front of our holding line, so I was among the first to emerge from the trees when we reached the edge of the forest. We entered the meadow, following our flames. Shadows gyrated across the field in long, twisting shapes. The sounds were the crackle, hiss, and pop of a campfire, but amplified. Another sound seemed to grow from the background, an ambient rumble that gained intensity until it was all I could hear. It pressed on us from every direction, bending the forest, creating wind, rattling the pines. And then I saw it: the megafire, approaching across the meadow.

I felt a stir of emotion that had become rare in my adult life. It touched me like an old scent—long absent, but unmistakably familiar. It was the feeling of standing before a Midwest stormfront as the calm breaks into a wall of cold air and washes over you. It was a feeling that reduced me to a sliver of existence, yet it somehow made me feel more alive than ever before. We stood on the edge of our planet's metabolism, where the biosphere is transformed into the atmosphere. And we were supposed to stop it.

I looked into the meadow. Aoki had parked his truck there, between our flames and the wild ones so he could monitor the activity of both. He was a silhouette in the night, standing between the fires. For a moment, I felt hope. Our flames seemed to have rallied in our defense, turning toward and galloping against the megafire. They gained momentum and grew, consuming the fuel between us before throwing themselves at the fire's head.

But then a sound like rain. The megafire was casting embers all around us. They fell across the road, starting little fires everywhere. I yelled, "Spot fire!" and ran toward a pocket of flames that scurried across the ground like a feral creature. I forgot my hoe, using my boots to kick pine needles away from its edge instead. Barba appeared on its other side with his Pulaski, a hybrid ax and hoe, digging a dirt line around the spot fire. His helmet was tilted to shield his skin from the heat. We worked together up the fire's opposite flanks, then met at its head, panting, and stepped back to stand on either side to ensure the flames stayed within our line.

I looked across the fire. Barba's face was covered in dirt, but he was composed, focused. There was no fear there, only a clinical calculation even as, through the forest, dozens of lights were growing on our side of the road, like a campfire the size of a forest—with us inside.

Looking back into the trees, I saw our crew was scattering, trying to encircle the embers even as the barrage intensified. It was impossible.

Barba finally raised his eyes to mine. "This is fucked."

I heard Edgar yell for everyone to get back to the vehicles. The megafire had crossed the meadow. It was time to retreat.

THE MOON WAS DEAD THAT NIGHT, THE SKY BLACK. IN THE DISTANCE, countless flames flickered in the mountains as if all the stars had fallen to earth. From those mountains came a high-pitched wail, the chainsaws of crews working through the night to hold our line. We would reengage with the megafire in the morning.

I lay in the dirt in the open air. The rest of the hotshots were scattered around me amid the bushes. We often slept like this. Aoki would lead our caravan far enough from the fire to avoid getting burned over as we slept, then he would park on the shoulder of the road. We would make our beds on whatever flat ground we could find. Hotshots call their beds graves.

I lay on my stomach to type my notes from the day into my phone. After the prescribed burn, with our own fire crackling behind us and the megafire growing around us, we had run to our buggies and sped down the road. The forest had begun to torch. I had felt the heat through the metal vehicle frame. Márlon's knuckles had been white on the steering wheel. "Relax, homie," I'd heard Red mutter from the passenger seat. "Slow down, just slow down. We're good."

Edgar had been driving a pickup truck that held our excess supplies. We followed him on roads that tunneled through the forest until we were out of the path of the megafire and could watch it from a granite knoll, burning toward a lake. The lake, I thought, would surely halt the fire, or at least slow it for the night. But as my crew gathered outside the buggies to watch the forest burn, I noticed a glow in the tops of the trees on the other side. Embers from the fire had already floated across the lake. It looked as if someone had hung lanterns from their highest boughs. The lanterns grew, then joined one another, rolling forward in red and blue light. The lake was soon engulfed. Its surface flickered as if burning from within.

I knew enough about wildfires at that point to understand that this was abnormal. Most wildfires follow something like a circadian rhythm, rising in the morning when the sun heats the land, then falling in the night when temperatures cool and humidity rises. Firefighters call this "overnight recovery." But this megafire was not following the usual patterns. It had overtaken our position as the sun set. Now it burned with such force that it created its own wind, carrying itself over all the barriers that would historically have slowed or stopped its spread.

I finished typing my notes and rolled onto my back, blinking against the green imprint that my screen had burned into my vision. When my eyes adjusted, I watched a bank of clouds move in a slow, uniform motion from one horizon to the next.

CHAPTER 7

A KICK WOKE ME IN THE DARKEST HOUR BEFORE DAWN. AOKI was a silhouette against the sky. When I moved, he drifted to the next body. I heard a thud, a grunt, and a rustle. The sounds were familiar. This is how we had woken every morning since spring.

It was the beginning of our fifth day fighting the Sugar Fire. We were sleep deprived from chasing spot fires the night before. We had not showered in weeks. Most of us were on our second of the two pairs of underwear we had room to carry. Our faces were black, our nostrils and lungs full of soot. The prescribed burn we'd lit that first night had failed to contain the megafire, and we had been overwhelmed at every other attempt. The fire now seemed unstoppable.

Before my mind could wake up, my body was moving, pulling on a shirt stiff with dried sweat. Scheer was already stumble-running through the dark toward Alpha buggy with unlaced boots and unzipped pants. Aoki had kick-started an informal race to prepare for the day. As the lead sawyer of Alpha Squad, Scheer tried to be the first.

These small routines had a way of softening the larger futility of our work. They created an internal order against the absurdity of our ineffective fight against the fire. My task, as the new guy, was to make sure everyone had coffee. Márlon had assured me this was the most important job on the crew. "Sometimes," Márlon said, "a warm cup of coffee is the only thing in the world we have to hold on to."

I hurried to the buggy and carried the coffee and camping stove to the saw compartment, where the lowered panel created a sort of working desk. Scheer was already there, tweaking his saw for the day's work. He moved to make room for me. Just a month ago, I might have wiped the panel clean of the grease and dirt and metal shavings that coated its surface. Now, I just started making coffee. We were all so filthy that it made no difference.

Edgar approached. I tensed reflexively. Whenever I entered his field of attention, whether I was sharpening my saw or boiling coffee, I worried that he might hover until he found a reason to publicly shame me. I normally tried to evade him, to stay out of sight and out of mind, but the saw compartment was next to Edgar's "world." As captain, he had a whole locker to himself in the side of the buggy.

The water began to boil, so I turned down the flame and stirred the coffee grounds. Edgar ignored me. He was shaving with an electric razor and seemed occupied by thoughts, tilting his neck against a mirror taped to the inside of his compartment door. Beside the mirror, as if to remind himself of his ethos, he had placed a black sticker with white words that read something to the effect of "I have no heart."

Sometimes, Edgar's performance of the alpha male was so ham-fisted that it seemed like a farce. Yet in spite of myself, I came to appreciate something in his attitude. In my other, softer, life at the university, where the only semblance of physical discomfort was occasional caffeine jitters, I had become accustomed to viewing the expression of feelings as therapeutic. By allowing the storms of inner life to cross the thresholds of our bodies, the thinking goes, we can smooth our internal waves and prevent them from manifesting in nasty ways—aggression, envy, and other destructive tendencies. The fireline, however, was forcing me to grapple with the idea that emotional expression may be more helpful in some contexts than others.

As I stirred the coffee and listened to the buzz of Edgar's razor, I felt exhaustion in my bones. It pressed against the back of my eyes and stabbed my joints. And I had been a hotshot for only a few months. I tried to imagine this discomfort stretched across the two decades of Edgar's career, with countless hours of grueling hikes, blistering heat, and the stress of keeping his colleagues alive.

To speak of these stressors would not make them go away. Nor would it change the nature of our work. It would only draw attention to the discomfort, making it more acute. For most people inhabiting this brute reality, learning to compartmentalize the pain is the first step toward surviving it. We all did this in our own ways. I intellectualized it. Scheer used humor. Others drank themselves into comfort whenever they had a chance. Edgar, for his part, built masculinity around himself like a fortress. The fortress was ugly, but he was safe inside.

On rare occasions, I would glimpse the man inside that fortress. Like the day we drove through a desert toward a brush fire and he gifted me an anthropology book, a history of work. The pages were smudged with dirt from his fingers. Perhaps he wanted me to read the book, or perhaps he wanted me to know that he was reading a book written by a Harvard scholar. But I did not need to open Edgar's book to grasp the acuity of the mind within that hard exterior. His intelligence came through when he crouched beside me on a sandy knoll where we watched a fire eat through a ravine. He pulled my attention to the sky, where wisps of cirrus clouds painted the upper atmosphere, then below, where a lonely cotton-ball cloud signaled an approaching weather front that would bring new winds and transform the flames. He carried my attention lower still, to the folds of the ravine, which would pull fire at different speeds when the winds came. Then even lower, to the shape of the ground, where he could read the root systems under our feet, which allowed him to predict how trees would fall when the fire burned through them. In a glance, Edgar could comprehend the world with an expertise surpassing that of any scholar. He shared this with a self-conscious reserve, perhaps because it betrayed the secret that he cared about us.

He expressed his care in strange ways. One day in the buggy, he turned off his music, turned back toward us, shades on, and quizzed each of us about our goals in life. The implicit meaning of his question was clear: he thought we could all be doing more. Scheer told him he wanted to create a tequila company. Edgar told another hotshot with a business degree to teach Scheer how to do that, to run the numbers. When the hotshot with the degree responded that it was only a piece of paper, that he spent most of college stoned, Edgar twitched. He looked personally affronted. When Drogo next responded sarcastically that he either wanted to be a lawyer for Big Tobacco or Big Oil, whichever paid more, Edgar seemed ready to give up. In a moment of courage, I asked what his career goals were. He twitched again, then turned around and turned up his music. "What are you talking about?" a hotshot whispered to me. "Edgar's the *captain*. He's made it."

Those moments of apprenticeship came frequently enough to leave us fiending for his approval, but they also preceded the worst outbursts. As if catching himself straying too far from his ethos, Edgar often snapped. "Do you know why I can do this?" he shouted at Scheer one day, after helping fix his chainsaw. "Because I have a brain!" Another time, after a hard day of work, he snarled at Drogo, "You're nothing. *Nothing.*" I wondered if Edgar needed to remind himself that, as captain, he was something.

After the coffee was done steeping, I approached Edgar carefully. I was supposed to deliver doses of caffeine along the chain of command, but Aoki had driven to base camp for a briefing, so Edgar came first. He had recently let his guard down, on the phone, during a rare moment of cell reception. I overheard him speaking to someone in a soft voice, promising that he would be home soon. He looked more tired than I felt, moving stiffly, with knotted shoulders and a tight jaw. I garnered his attention as I would an injured animal, wary that he would notice his vulnerability and assert himself. Instead, he waved me away, telling me to come back with the coffee after the boys had their fill.

By the time Aoki's truck crunched into our parking area, the sun had just blazed over the horizon. We gathered around Aoki as he fixed the fire map to the side of the buggy with magnets. The map showed the entirety of the fire, which was nearing one hundred thousand acres. The edges were traced with colors showing where crews were building lines. The fire was zero percent contained, which meant that none of the lines had held.

He pointed to a critical segment of the fire, a ridge where, if the fire tipped over, it would spread like water from an escaped dam. Today, we were going to cut line around that segment. The stakes were high: hundreds of people lived on the other side. To ensure that we could stop the fire, we would be "going direct."

"Going direct" meant we would be working directly on the edge of the flames. We had been assigned to Q Division, the steepest, most inhospitable terrain in the region. These assignments are exclusively reserved for hotshots. This was what hotshots lived for. It was what we were trained for. It was how we maintained our reputation as the best. Nevertheless, I felt dread settle in my stomach. This would be my first time going direct since training, and Edgar's words were still branded in my mind. *You're not ready yet.*

After reviewing the map, Aoki held up a sheaf of papers, the daily briefing. He squinted at the papers through reading glasses as if surveying a familiar menu. "Now listen up. Today we have unprecedented heat, historically low fuel moisture, historic fire conditions." He ticked through a litany of signs of climate change in a bored voice. "Expect extreme fire activity . . ."

He lowered the papers. "Look, fellas," he said. "If I was a boxer, I'd throw in the towel. I'm tired of getting my ass kicked." Then he grinned, surveying us. We were as puffy-eyed and soot-faced as nineteenth-century coal miners.

"You know something that most people don't understand about our job?" Aoki asked. "It's that we don't really fight fire. We read the conditions. We wait for an opportunity. When the fire lays down, we get around it. Then, when it stands up, it has nowhere else to go."

For us, the meaning behind his words was clear. The best conditions for firefighting are when smoke looks like an anvil several hundred feet above the ground. The smoke forms that shape when the atmosphere is layered, creating a cap that limits available oxygen and, by extension, the severity of the fire. The next best conditions are fog low over the earth, which makes the fire gasp for breath, but also limits visibility and precludes air support. Today, the dawn was deep blue. There were no boundaries to the upward expansion of the fire. It could feed from all the oxygen in the atmosphere.

We needed to hurry. Before the heat of the day dried the brush, we needed to carve a line along the fire's flank, between wilderness and society. Maybe, if we were lucky, tomorrow's conditions would allow us to approach the head of the wildfire and cut it off.

BEFORE HOTSHOTS BEGAN CARVING LINES THROUGH AMERICA'S LAND-scapes, other lines were drawn on maps that covered the carpets of the

White House. The year was 1907. Though the U.S. Forest Service was only two years old, it appeared that it might die in its infancy, strangled by legislation that, in one week, was poised to remove the president's authority to conserve land in much of the West. Just before the bill was signed into law, President Theodore Roosevelt crawled over these maps on his hands and knees, intent on protecting in one night all the lands that were about to be removed from his jurisdictional authority.

"Oh, this is bully!" Roosevelt snorted through his mustache. Beside him, on the ground, was Gifford Pinchot, his lanky forestry chief. "Have you put in the North Fork of the Flathead?" the president asked. "Up there I once saw the biggest herd of black-tailed deer." Roosevelt and Pinchot fed on each other's energy, enlivened by landscapes they imagined as primordial wilderness.

The idea of American wilderness, according to environmental historian William Cronon, was invented gradually. To imagine the land as wild, American colonists first needed to remove Indigenous people from that land. Then, to imagine the frontiers as "untouched," the American public needed to erase Indigenous histories of fire and land management. Finally, with landscapes brutally emptied of Indigenous people and their histories, the dominant figures of colonial society filled the void with their deepest core values: profit, on one hand, and conservation, on the other. By the time Roosevelt took office as president at the dawn of the twentieth century, the landscapes of the American West had become a living canvas upon which those in power projected their cosmic dramas.

For the president, the wilderness that filled his maps was more than scribbles on parchment. In his youth, Roosevelt had taken a sojourn across the West, sleeping on hard ground beneath stars, galloping across unbroken vistas, and reveling in landscapes that had been shaped by Indigenous people, but that he imagined as primordial wilderness. This land was like spiritual medicine for Roosevelt, a tonic that mended, then fortified, a heart broken by the untimely death of his first wife. "I owe more than I can ever express to the West," he wrote some years later.

While Roosevelt imagined the forests as cathedrals of nature, others saw potential for lucrative profits. Following the genocide of Indigenous people, the lands of the West had been opened for exploitation. "Never again," historian Frederick Jackson Turner proclaimed in 1893, "will such gifts of free land offer themselves" to the American people.

But "American people" really meant the industrialists who ruled the rest of the economy—timber barons, mining magnates, oil tycoons, and railroad monopolists. These industrialists had already leveled the forests of Maine and Michigan, churned through the Appalachian hardwoods, and now had their sights set on the white pines of the northern Rockies and the redwoods of the California coast. Roosevelt described these industrialists as "the most dangerous members of the criminal class, the malefactors of great wealth."

By the time Roosevelt took office as president in 1901, conserving the wilderness of the American West was a centerpiece of his agenda. In his first message to Congress, he declared, "The forest reserves should be set apart forever for the use and benefit of our people as a whole and not sacrificed to the shortsighted greed of a few." When off script, Roosevelt was more direct. He was reportedly prone to fits of passion in which he would bellow, "I am against the man who skins the land!"

At the time, however, the idea that corporations should not "skin the land" was a novel concept, akin to sacrilege. Many considered the wild to be a vile, godforsaken place. The wilderness was where God banished Adam and Eve and where Jesus was tempted by the devil. It represented moral confusion and despair. If land was not being exploited for wealth, it was considered a wasteland—quite literally, "wasted land." Conveniently for American profiteers, waste was sin. By churning wilderness into profit, therefore, the industrialists were doing more than looking after their own coffers—they were the vanguard of a moral order. Just as, in their minds, Western civilization had tamed the sinful nature of humankind, so were industrialists obliged to tame that same nature in the external world by chopping down the wild and sowing it with order. To imply that American industry was wrongfully destroying the wilderness was to contradict deeply held beliefs about the special covenant between white Christians and God.

So there was a cosmic upheaval on the Senate floor as Roosevelt expanded the national forest system, with the religious angst of racial and economic supremacy refracted through the language of the Constitution. Senator Weldon Heyburn of Idaho implied that the Forest Service was committing treason by violating state's rights. A newspaper owned by Senator William A. Clark, a notoriously corrupt Montana businessman, warned that conservationist ideals were "infesting the West." Later, the secretary of the interior, Richard Achilles Ballinger, was more direct in his critique, lamenting that the Forest Service was hindering the march of progress. "In my opinion," he said, "the proper course is to divide it [the wild] up among the big corporations."

The task, for Roosevelt, was to flip the script of the wilderness. If capitalists camouflaged their economic interests with a moral imperative, the president must do the same for conservation. Roosevelt toured the West, barking speeches from the rim of the Grand Canyon and the granite domes of Yosemite, but he also needed someone to enforce this image on the ground. For this task, he chose Gifford Pinchot, a rail-thin forester with the distant eyes of a missionary.

On the surface, Pinchot was not a logical choice for the endeavor. His place in the new American aristocracy had been secured for him by his grandfather, a timber baron who made a fortune deforesting Pennsylvania. In his youth, Pinchot seemed poised to pursue his family trade when he traveled to France for his studies. At the time, forestry was still a young scientific enterprise, having emerged in northwest Europe to provide the raw materials for the Industrial Revolution— the railroad tracks, mining fortifications, and factory walls that allowed it to surge ahead with unprecedented force. Foresters, in Europe, were not trained to enhance the biological diversity that sustains food systems and reduces the risk of catastrophic wildfire, but to make trees orderly so their board feet—the amount of lumber trees hold— could be calculated to maximize harvests. The forests of France, Pinchot wrote, resembled plantations, grown like a crop.

Rather than propelling Pinchot into a career of industrial forestry, however, his experience in Europe left him disturbed. Forests and foresters seemed to have almost nothing to do with one another. When he returned to America and traveled across the West, the memories of domesticated timber reserves created a haunting juxtaposition in his mind. Set against a Rocky Mountain peak where snow lashed his face and forests darkened the horizons, Europe's timber plantations felt like a warning. To plunder America's forests, for Pinchot, was not only shortsighted, but the pinnacle of moral depravity. He joined Roosevelt's crusade with the zeal of a convert, intent on shifting the idea of wilderness in the public imagination and, in doing so, preserve the future of the forests themselves.

Yet even as Pinchot rejected the extractive premise of industrial forestry, he internalized the industry's hallmark fear of fire. French industrialists believed that flames served no useful purpose, but only existed as a threat to timber harvests. When Pinchot gained control of the Forest Service, he needed to do more than quarrel about economics. He needed to capture the notion that wilderness was a battleground between good and evil and turn it to his own purposes. To do this, Pinchot ascribed a contradictory higher purpose to the Forest Service: to vanquish natural fires from the wilderness in order to conserve nature itself. "Of all the foes which attack the woodlands of North America," Pinchot wrote, "no other is so terrible as fire."

Pinchot handpicked the staff for his new agency, choosing individuals for whom conservation held religious appeal. "The Chief," wrote one of his first foresters, "taught us to see God in nature." He found an ally in John Muir, who had found God in Yosemite Valley. Muir particularly loved Tenaya Lake, which had recently been named by the commander of a California battalion. After burning Ahwahneechee villages in Yosemite, the commander captured their leader, who was named Tenaya. "I called Tenaya up to us, and told him that we had given his name to the lake and river," the commander said, according to a militiaman who recorded the events. "At first he seemed unable to comprehend our purpose, and pointing to the group of glistening peaks, near the head of the lake, said 'It already has a name.'" The new name was a mockery, to remind the tribes that they would never be allowed to return. Muir instilled this mockery with the myth of untouched wilderness, where fires, and the people who lit them, were intruders.

For Muir, along with most conservationists at the time, fires posed a significant threat to the wilderness he sought to protect. Though he was mostly concerned with preventing industrial exploitation, he referred to fire as "the master scourge." Writing about the sequoias in 1869, Muir argued that "only fire threatens the existence of these noblest of God's trees." He advocated for fire prevention and suppression to be incorporated into federal plans to protect America's forests.

Roosevelt and Pinchot lent Muir their ears. In the first years of the agency, using nothing but horses and shovels, the Forest Service eliminated fires from all but a fraction of a percent of public land, championing a 97 percent rate of suppression. While scholars have attributed this early success to a lack of fuel buildup thanks to millennia of active burning and land management by Indigenous people,

President Roosevelt saw fire suppression as proof of the efficacy of the agency. "It had a great task before it," Roosevelt wrote to Pinchot in 1906, "and the Forest Service has proved that forest fires can be controlled."

The industrialists, and their cronies in Congress, weren't convinced. When Roosevelt left office in 1909, capitalists moved in for the kill, starving the agency of funds, reducing its ability to advertise itself, and waging a stream of lawsuits. Roosevelt had left Pinchot with the charge of protecting 230 million acres of land, an area equivalent to the Eastern Seaboard from Maine to Florida, but Pinchot was stripped of the resources he needed to carry out this task.

In 1910, a lightning storm produced a firestorm in the mountains of Idaho, with flames riding winds of eighty miles an hour to incinerate three million acres in a matter of days. Seventy-eight firefighters were killed battling the blaze. Pinchot capitalized on the opportunity to frame them not only as heroes, but as victims of a greedy industry that had crippled the ability of the Forest Service to carry out its central task of fire suppression. As headlines across the nation regaled the public with images of shelled towns and scorched earth, Pinchot pressed his message into the fray. "Today, we understand forest fires are wholly within the control of men," he concluded. Those men simply needed more funding.

And the funding came. Ten months later, Congress doubled the budget of the Forest Service. But the success of the conservation crusade held a tragic irony. In their fight to protect the land from their own society, Roosevelt and Pinchot had elevated the sacred appeal of the wilderness to the highest levels of government. Within this imagined wilderness, neither people nor fire belonged. As a result, the Forest Service institutionalized the same exclusions that formed the core of the colonial encounter in the American West, ensuring the degradation of the very forests they sought to conserve.

Once drawn, those lines Roosevelt and Pinchot had created on their maps in the White House needed to be inscribed on the land itself. The lines needed to be patrolled. They needed to be defended. This would be a task of bodies and blood.

"From that time on," Pinchot later wrote, "it was *fight, fight, fight.*"

※

THE HOTSHOTS AND I WERE HIGH ON A GRANITE MOUNTAIN. I FELT LIKE we were fighting for our lives. Several miles to the north, the head of the fire loomed like a thundercloud. We had found the flank of the Sugar Fire where the flames lapped low to the ground, but they were climbing higher as the sun sapped humidity from the air and moisture from the brush. We hoped our line would prevent this edge of the fire from spilling over the ridge and taking off.

Márlon, face striped with dirt and sweat, stood on a boulder above our crew, issuing commands to Chinook helicopters that dropped water on the flames. The helicopters kept the flames low so we could get close. Axel was higher up, on a mountain summit, monitoring the weather and reporting any changes in wind patterns on the radio so we would know if we needed to flee.

I wielded a chainsaw, expanding the swath carved by Scheer. Scheer moved gracefully, slicing and slashing with stamina that never seemed to wane.

Now, Edgar was encouraging us to work even faster than usual because we were in danger. Hotshots are trained to always keep "one foot in the black," staying on the edge of the area that has already burned so as to be able to escape into it if the fire flares up. But on the slope, we were surrounded by brush with the fire below us. Fire burns quickest uphill, heating the vegetation above, so it was only a matter of time before the vegetation dried out, the wind picked up, and the flames roared to life. Márlon's helicopters were holding the inferno at bay, but the burnover seemed imminent.

I stuck close to Scheer, trying to find a rhythm of breath and movement. After two hours of hacking, however, my muscles began to cramp, and my mind raced. *I was too focused on coffee this morning. I didn't sharpen the chain. It's still dull from last night.* Then my saw sputtered and died.

On the fireline, problems cascade. A dull chain heats the bar, burns the oil, and blows out the spark plugs so that the machine can't get the fuel it needs to run. When one saw fails, everyone else is forced to cut more, slowing down the crew. And when the crew slows down, everyone is at a higher risk of death.

I pulled the ignition cord. The engine purred. I cut a few more bushes. It died again. This time, I pulled the cord a dozen times, cursing and sweating. My forearms seized up. I lost control of my hands. The stress built as the gap grew between me and the other sawyers. My puller was cussing me out, which didn't help. Red had caught up to me and was standing behind me, waiting. The rest of the team was behind him. We were above the fire, and I was slowing the crew. This was how hotshots died.

After what seemed like an eternity, I emerged, alone, onto a slope where a rockslide had wiped out all the vegetation. Edgar was waiting, arms folded. He marched over and held out his hand. "Don't take this personally, Thomas," he said. The other sawyers sharpened their chains, adjusted their packs. No one met my eye except Edgar. His face was earnest. He looked like he regretted what he needed to say. This wasn't about me. I knew what was coming, and I knew that, this time, it really wasn't personal.

"I need to look out for the safety of the crew," Edgar said, so quietly that only I could hear.

I couldn't meet his eyes. I'd felt the pressure of his scrutiny since training, and now I had failed. Another sawyer had told me that their saw would need to be pried out of their cold, dead hands before they gave it up. It was a loss of status, a relinquishing of honor. But with the fire below us, the danger was all too real.

I held out my chainsaw. Edgar took it from me.

Edgar surveyed the group. All the pullers stood a little taller, hoping he would hand them the machine. His gaze settled on Drogo, who stood the tallest. Drogo winked at me. I gave Drogo my chainsaw repair kit. He gave me the fuel canisters he carried for his sawyer. From then on, I would be Scheer's puller. Everyone fired up the chainsaws, and we got back to work.

At first, any shame I felt was accompanied by the physical relief of relinquishing the twenty-five-pound saw. That relief was short-lived. Scheer dove into the brush, tunneling forward with fresh speed. As I wrestled falling bushes and dodged the teeth of Scheer's saw, I began to realize that "pulling" is the most demanding position on the crew. Thorns and twigs shredded my leather gloves and slid under the skin of my forearms until the flesh became a flayed web of cuts and blood. The pain was acute. More than once, hefting bushes upslope, I tripped on roots and went rolling down toward the flames until I caught myself and scrambled back to the crew. But eventually, the pain faded, and I entered a new mental terrain.

I've come to imagine this terrain as a dark mountain. At the top lies total physical breakdown, but in the ascent, I crest a series of false peaks. At each false peak, I feel I will collapse, but then I see the mountain rising higher still, and my body keeps moving, climbing into new levels of dissociation. Time distorts, sensations fade, and instinct and movement take the place of rational thought.

I couldn't allow rational thought to flee entirely, however. The job required an incredible attention to detail. Scheer was cutting vegetation directly along the fire, which meant some bushes were burning and others were not. As I caught the branches, I needed to sort them, throwing smoldering wood downslope, into the fire, and cold wood upslope, into the green. If I tossed a single branch to the wrong side, I could start a fire above us, compromising our mission, burning people's homes, and risking our lives. I couldn't easily see if the blackened branches held any residual heat so, to keep up with Scheer, I pressed the wood against the exposed skin on my neck. If I felt my skin burn, I knew the wood was hot.

The world spun. I found myself on my back, soaking wet, knocked flat by around a thousand gallons of water that now rushed downslope. Márlon shouted an apology from his position on a boulder above us. He had misguided the helicopter. I could see a line of other helicopters approaching, balloons of water dangling from cables attached to their bellies. I noticed that the shadows of the mountains had deepened. Somehow, we had been working for hours, though it felt like only minutes had passed. Scheer saw that I had fallen and came to stand over me. He stalled his saw and offered his hand, pulling me to my feet.

"You ready?" he asked, eyes wild. The water had brought me back to my senses. I could now hear the howl of chainsaws approaching from down the mountain. Another hotshot crew was close. When we met, it would complete our line around this segment of the fire.

Before I could process what was happening, Red was hollering from behind us. "Tie it in, fellas! Show 'em who LP is!"

Our own saws howled back to life, and we piled forward with renewed force, branches flying, sawdust spraying, bodies ducking and weaving and cutting and throwing. "Tying in line" with another crew is like meeting a rival team on the sports field. It is a point of pride to tuck away all the fatigue and meet them with as much force as possible. I had heard stories of rival hotshot sawyers fighting to cut the last bush until their chainsaws clashed and sent sparks flying. From the outside, the spectacle may seem buffoonish, a twisted ethic of machismo that whips tired, exploited bodies into harder, faster work with no tangible reward. But in the moment, the adrenaline was all-consuming.

I could see bushes moving ahead, with white helmets bobbing beneath. Scheer ran the saw at full throttle. I pulled away one final bush to reveal a huge and unfamiliar man, grizzled and sweaty and hefting a chainsaw far larger than our own. Scheer stepped forward, face set in a silent grimace. I moved aside as the sawyers circled each other like swordsmen in a duel, cutting every last branch they could find until none were left and they faced each other with heaving chests and sweat pouring through the grime. They revved their saws. At this point, both crews had gathered around. Scheer seemed to realize that he had been caught in some strange ritual; he waited until the other sawyer silenced his chainsaw before he turned off his own.

We faced the other crew until the silence became awkward. Then we shuffled past them. We walked down the line they had cut, inspecting it, judging the quality, making noises of disapproval until we were out of sight and could sit without exhibiting weakness. The line connected to a knobbed granite spine that jutted across the landscape. The stones created a natural fuel break that completed our barrier around the fire.

I sat on a rock and looked over a black skeleton forest turning gold with the sunset. My whole body vibrated. Scheer and Barba crouched down beside me.

"Eyyy, you earned puller's wrists," Barba said, smiling. He was the lead puller, and he rolled up his sleeves to show me that his tattoos were lacerated, like his arms had been locked in a box of feral cats. "You killed it today, Thomas."

"Yeah, you did," Scheer added. "You killed it."

I realized they were trying to comfort me for having lost the chainsaw. I appreciated their care, but I was too tired to speak, much less feel shame or pride. In a day's work, we had created a line around one corner of the fire. Looking in the other direction, I saw a sunset-tinged thundercloud and felt a moment of hope, thinking rain might possibly extinguish the rest. Then I realized my mistake. The bulbous shape came from across the mountain range where the wildfire was ejecting the remains of the whole forest into the sky, creating a pyrocumulus column that tumbled as high as the stratosphere.

CHAPTER 8

WHEN AOKI ASKED IF I HAD EVER BEEN INSIDE A FOREST FIRE, I thought I heard him wrong. The day before, I had cut line along the fire's flank from sunrise to sunset. The evening before that, I had chased spot fires through torching trees that turned the forest into a blaze of light. As we approached the final day of our two-week assignment, I felt like I had been living within wildfires for months.

"Inside a forest fire?" I repeated as I sat beside him in the passenger seat of his parked utility truck. "Like we are now?" I tried to keep my voice level despite the pine trees exploding several hundred feet upslope.

With the flank of the fire secured, Aoki had chosen Scheer to accompany him and Axel on a scouting mission to formulate the next stage of the containment plan. As Scheer's new puller, I was part of the entourage. Just minutes before, we had retreated to Aoki's truck from a nearby ridge when his voice crackled on the radio to tell us our position was no longer safe. So I was puzzled by Aoki's languid demeanor as he propped his booted feet on the open window of the open truck door and raised an eyebrow at me.

"No," he replied. "I mean *inside* a forest fire."

"I guess not."

Aoki grunted and looked back to the smoking forest. His hair was tied up so it could be tucked under his helmet, making his face seem longer, the lines deeper. Scheer was outside, trying to find the keys to a different truck that needed to be evacuated before the fire blew over our position. Axel had taken a seat in the driver cage of a bulldozer and was trying to teach himself to drive it. The machine lurched and jolted as he pulled various levers.

Until now, my view of fire operations had been confined to a cloud of sawdust and crush of brush. With Aoki, this was my first chance to understand the big picture, to see how hotshot leaders plan their attacks and coordinate with central command to deploy the airplanes, helicopters, and crews that cut and plow and loop a giant lasso around the fire's head. I had assumed Aoki would create his plans far from the fire, but the forest was beginning to shake around us. I prodded him.

"We'd be fine here, right?" I asked, referring to the logging road we were parked on.

"Hell no." He seemed to find the idea amusing. "I mean we *might* survive here, maybe, but it wouldn't be fun. It would be tragic, actually. You'd probably quit. I've cooked inside a vehicle once, and trust me, it's not fun. You're like a turkey in the oven."

I looked at his hands on the steering wheel. They were discolored from skin grafts he had received after being overtaken by flames several years before. When he knew he was trapped, Aoki sought shelter in a truck, but the first wave of superheated gas cracked the windshield and pressed against it until it groaned and threatened to shatter. He leapt from the vehicle, using his hands to protect his face, holding his breath to prevent his lungs from melting, and running through the inferno. He fell, leaving the skin of his palms behind on the searing pavement. He fell again, raw sinew pressed against the ground. When he finally stumbled from the flames, flesh hung from his wrists like loose gloves. "It's a pain that never goes away," he told me, flexing his fingers. "But that day, I learned exactly how much heat I could take."

This recollection did not reassure me as I sat in Aoki's truck with the sounds of the megafire growing louder around us. I paused to remind myself that Aoki was probably the most experienced hotshot on this fire. The inside of his truck whirred like a spaceship, with spinning dials, crackling radio traffic, and a digital map of the entire region. The fire was represented on the map by a blob of red overlaying topographical lines, with solid black edges marking where crews had successfully contained its expansion. One line was the ridge where, the day prior, our crew had spent twelve hours carving through brush. It looked tiny.

Today, Aoki had brought us to this particular ridge because, on the map, it was well outside the red blob. On the map, the ridge looked promising, with logging roads tracing contours that bent around the arc of the fire. If the blob had stayed where it was on the map, Aoki would have been able to organize an attack, bringing bulldozers up the roads to push wide swaths of dirt, hotshot crews to burn everything on the other side of that dirt, and less experienced crews to spread out behind us to watch for spot fires. This was called a "big box" operation: often, the only way to contain a megafire is to create an enormous box around its head, creating fuel breaks along the ridges for miles around and burning everything between. But this fire was escaping the planned box before Aoki could even set it up. He was already squinting at the map to find new ridges.

A furtive voice came through the radio traffic. Aoki turned up the volume. "You hear that?"

"I missed it." There were half a dozen robotic voices squeaking through at any given moment. Aoki liked to claim that we'd know as much about the fire as him if we just listened to the radio, but how he made sense of the noise baffled me. He absorbed the voices while he ate, while he worked, probably while he napped, and from their static he could somehow piece together everything everyone was doing on the fire all the time.

"There was a request for a medical air evacuation. Listen."

A male voice came on the radio, reporting a broken femur. A firefighter had fallen from a cliff. His crew was attempting to carry him to a knoll, where they would cut a landing pad for a helicopter so he could be carried to a hospital.

"Know where they're at?" Aoki asked.

"No."

"They're on the line you cut yesterday."

Aoki grimaced, showing coffee-stained teeth, but I sensed a touch of pride in his voice. On a fire of this size, with multiple divisions often staffed by hundreds of people, competition inevitably emerges. Hotshots are the tip of the spear, cutting through brush along the fire's edge, followed by engine crews who "plumb" the firelines with hoses. Those firelines are then occupied by less experienced crews who patrol the line to make sure the fire doesn't cross. The least esteemed task falls to those who "mop up" after the edge is secure, venturing into cold ashes to extinguish any persistent embers. Hotshots like to refer to people with hoses as "engine slugs" and beginner crews as "mop shots." The injury report, while tragic, also conveyed a message: other crews hurt themselves just trying to patrol the lines we built.

Another tree popped near the road, created a swarm of embers. The embers settled into the canopy, winking at us. The fire was gaining momentum.

"Does that concern you?" I asked Aoki, bringing his attention back to the fire, trying to stay cool.

"Well, yeah! That's why I'm getting everyone the hell out of here."

"What are we waiting for?"

"Helicopters."

"Helicopters?"

"The fire's beneath us now. We need them to dump some buckets so it doesn't rush up at us while we're driving out."

Outside, Scheer had found the keys to the truck and looked ready to go. Axel had the bulldozer under control. The embers weren't winking anymore. They were waving at us over the treetops. I settled in my seat beside Aoki and waited for the helicopters to come.

AN HOUR BEFORE, WHILE AOKI WAS IN HIS TRUCK MONITORING THE DI-rection of the fire column, Scheer and I worked on the ridge above, where Axel had taken us scouting to see if the logging road could be used in Aoki's big box plan. Though we were a mile away from the main blaze, we had found spot fires glowing all around the forest floor. While Axel called a bulldozer to push dirt around the bigger patches of flames, Scheer and I contained the smaller ones by kicking away pine needles and cutting low branches off trees so the flames couldn't climb into the canopy. I no longer found the spot fires frightening. So long as they weren't scurrying toward me, they seemed as normal as puddles after a rain.

I had been paired with Scheer for only a few days, but I found him to be surprisingly good company, mainly because he was unabashedly weird. While we worked, he told me that he used to mop floors in yoga studios in exchange for free lessons. He went to art school to be a surf photographer. He dropped out after partying too hard. On the fireline, I noticed, he often talked to himself, often to berate himself for dropping out and partying too hard. During our breaks, he wrote sad poetry and raunchy songs. He also collected dead things for his girlfriend, who was a model in Los Angeles. When I later visited his trailer back in Santa Barbara, I found it festooned with skulls and ribs and feathers. The trailer teetered on a slope where every gust of wind threated to hurl it into a chasm, which seemed like a metaphor for the frenetic energy of his life. He was the kind of guy you just wanted to help. I eventually began making a point to bring him snacks and vitamins and medications. Scheer wanted to help me too.

"You should really learn to surf," he advised me, cutting a branch over one of the spot fires.

"I've tried it," I said, catching the branch before it fell into the embers. "Haven't gotten into it."

"You should. You'd learn to move better so you wouldn't look like such a kook on the saw."

He didn't mean it as an insult. This was sage advice from Scheer. I asked if he thought yoga would help me move like a sawyer.

He paused. "No, definitely not."

We heard Axel before we saw him, leading a bulldozer amid a snap and crunch of trees. Hotshots usually guide bulldozer operators to make sure the operators are pushing firelines in the right locations and to ensure they don't drive off a cliff or flip the machine on too steep a slope. Scheer and I hiked up to the road to get out of the way.

At first glance, watching him approach through the trees, Axel seemed like an older version of Scheer, with strong arms and shoulders honed by surfing and a voice that could drone like Matthew McConaughey's. But their difference lay in their gaze. While Scheer had the doe eyes of a romantic, Axel's were sharp, blue, and focused. His laugh was bubbly, always close to the surface. When he saw us by the spot fires, his face crinkled with delight, as if this were a chance encounter with old friends.

"Whoa," Axel said, coming to stand beside us while the bulldozer attacked a patch of flames. "I didn't see that before." He wasn't looking at the spot fire, but to the sky, where a red and black column towered over us.

I noticed Axel wasn't breathing hard, even though he had just hiked up a mountain. Axel had nearly as much hotshot experience as Edgar and commanded a different form of respect. His fitness was part of it—at thirty-six years old, he could out-hike anyone on the crew, including Aoki. But his leadership stemmed from his approachability. He viewed hierarchy as dangerous because he wanted Bravo Squad to feel comfortable pointing out hazards he may have missed. On the ridge, eyeing the fire column with suspicion, he produced organic bamboo papers to roll a cigarette.

"We're in a lot of Watch Out Situations right now, guys. You know which ones?"

Scheer scrunched his face like he had been hit by a pop quiz. "Um, unburned fuel between you and the fire?"

"Escape routes and safety zones not identified," I added.

"Yeah, that's right." Axel lit the cigarette. "Also, wind increasing speed and changing direction. But don't worry, guys, I found some bitchin' black down the hill in case we need to run."

We stood, passing the cigarette, watching the bulldozer smash trees, churn soil, and pump fumes as it circled the spot fire. I loved the bulldozer in spite of myself. All I could think was that I wouldn't have to do that work.

"Having dozers around is pretty nice," I offered.

Scheer nodded. "Hotshots' best friends."

Axel shrugged. "Makes our job easier. Environmental footprint sucks though." Like so much of our work, the footprint sucked, but it was part of the job. A bulldozer clears more vegetation with less discrimination and with more soil disturbance than a hotshot with a chainsaw, all while producing a cloud of diesel fumes. If Axel wanted to someday be promoted to Edgar's position as captain, he needed to qualify as a Dozer Boss, which meant he needed to lead bulldozers around fires. The footprint was the cost of advancing in his career, as well as battling the fire.

"Do you ever wonder why we're out here doing this?" I asked. "Putting fires out?"

"Oh yeah, all the time," Axel said. "It's complicated, you know? It gets political with all the stakeholders. Like, this is a logging forest. And the loggers definitely don't want to see these trees burn."

"This is a logging forest?" I was surprised. After spending the past few weeks amid charred trunks with skeleton limbs, I had been enjoying the green canopy. I hadn't thought of the history of this particular forest. "How can you tell?"

"I mean, the signs are everywhere. Look how I led the bulldozer up here. See that kinda flat part crossing the slope? That's an old logging road, where they drug out the original trees. Found some rusted tin cans too. Once you know where to look, the signs are everywhere."

The observation hit me like vertigo. Now that I knew where to look, I couldn't stop seeing it. While an old-growth forest would hold a mosaic of different trees, young and old, clumped and scattered, the trees around me were all the same species, the same age, and crowded. I suddenly felt uncomfortable. My own senses had been deceiving me. We weren't on a wooded hill in an old forest, but in a pine plantation.

This forest would have looked entirely different in the years between the beginning of fire suppression and the dawn of industrial logging. Fire suppression would have left its mark, with bushes growing under the small trees, small trees clustering around old trees, and the seeds of the old trees lying dormant under a deep litter of pine needles. But the forest would have been diverse, and that diversity would have served as a natural check on the spread of wildfire. A meadow, a gap in the trees, a carpet of ferns, a stand of old growth—all of these would make wildfires burn in patches. Flames would leap into the canopy in some places, but crawl slowly across the ground elsewhere. Variety in forests creates variety in flame types, and this ensures that some plants of each species can survive, reseed, and regrow. But this forest was different. It was not just a landscape starved of fire, but a new crop of trees planted on the tomb of the old forest.

A century ago, despite the premonitions of Gifford Pinchot, the U.S. Department of Agriculture was given jurisdiction over the newly created national forests. The subtext of this association was clear: forests were a crop that the Forest Service would manage. For one shining decade, the association of forests with agriculture was in name only. During the reign of Pinchot, a central task of the Forest Service was to protect the land from industrial excess. While fire was the rallying cry that lent the agency popular support, fire suppression was a means toward an end, not the end itself.

This quickly changed. When Pinchot was fired by the following presidential administration, the logging industry saw an opportunity to influence his successors; in particular, William Greeley, a baby-faced man with circular spectacles. Through Greeley, the logging industry turned the Forest Service to its advantage. The son of a pastor, Greeley called himself a "forest missionary." When he was caught in the Great Fire of 1910, which burned three million acres in the Northern Rockies, he felt he had been defeated by Satan himself. "I was spurred on by vivid memories of blazing canyons and smoking ruins in little settlements and rows of canvas-wrapped bodies," Greeley said. "This fire woke everyone up."

People had awoken, according to Greeley, to the notion that forests would inevitably burn unless the Forest Service scaled up their efforts to prevent fire. The logging industry offered a solution to this problem. If wildfires were the greatest threat to forests, why not cut those forests to prevent fires? The agency would build the roads, scout the timber, and protect the lumber. In exchange, logging barons would put the weight of their congressional influence behind the Forest Service. Greeley loved the idea. "The public," he wrote in 1920, "has the right to expect the co-operation of the large western timber owners."

By the 1920s, the Forest Service depended upon the logging industry for its fiscal survival, and the logging industry depended upon the Forest Service to protect its profits from fire. Greeley summarily quashed all fire management proposals advocating for anything other than total suppression. He denigrated the idea that some forests *need* fire as "an insidious doctrine" practiced by "the noble redskin." In an article published in *The Timberman*, an industry magazine, Greeley wrote that prescribed burning had represented "a loss to the forest resources of California today which we can safely put at 37 billion feet of standing timber, with a value of probably $75,000,000." If controlled burning was not stopped, he continued, "the end is total destruction."

To an ecologist today, aware that certain kinds of fire uphold the integrity of most California landscapes, Greeley's proclamations were clearly wrongheaded. But Greeley's logic was shaped by his political situation. With his agency captured by the logging industry, he judged the forest by its commercial value. Fire threatened corporate profits, so fire had to be stamped out.

Following Greeley, the Forest Service cut roads deep into the wild. Lumbermen used these roads, using chainsaws to slash the biggest trees, burn the rest, and haul lumber through the muck. In days, entire mountain flanks were transformed into mud and stumps. In months, a logging company could denude an entire range. Within decades, around 70 percent of California's old-growth forests were gone. The same agency that had been created to conserve the land had become a tax-subsidized vessel for extractive corporate forestry. "Absolute devastation," Pinchot wrote in his diary later in life, visiting the forests he had tried to protect. "Mostly a clean sweep, taking everything."

The mud and stumps left behind after a clear-cut don't generate profit, so the logging industry—and the Forest Service—promptly set about refilling the landscape with whichever trees could be sold at the highest price. They contracted airplanes and helicopters to buzz over the open slopes, spewing thick clouds of homogenous seeds. In some places those seeds were Douglas firs, used to build furniture and cabinets. On the ridge where we stood, those seeds were lodgepole pines, used to make telephone poles, fence posts, and railroad ties.

By the time Greeley retired from the Forest Service to serve as an executive for the West Coast Lumbermen's Association, he could claim that his tenure had constituted a resounding success. Fire suppression had allowed for the unprecedented extraction of commercial value from public land, while ensuring not just the survival but the expansion of the Forest Service. All "merchantable stumpage" was "fully protected," Greeley wrote, with "billions of acres of National Forest pine lands demonstrating the results of fifteen years of successful protection from ground fires." As evidence, he pointed to the new forest plantations beginning to shroud the mountains, where "the actual growth of timber has increased several times over what it was during the days of periodic fire."

In the forest with Axel and Scheer, I could see the truth of that statement firsthand. In terms of board feet, this forest was optimal. But in terms of fire potential, it would be difficult to build a more explosive environment. The lodgepole pines were crowded by white firs, a shade-tolerant species that had proliferated in the absence of fire and could easily carry flames into the canopy. The slope across from us bore the legacy of a clear-cut: mostly the same species, mostly the same age, with none of the heterogeneity that would have historically slowed the spread of fire. On average, landscapes created by corporate forestry hold approximately seven times the density of those managed with fire. Each of these factors—the homogeneity of the trees' age, the standardization of tree species, and their density—has combined with climate change to transform forests into tinderboxes. These forests are primed to burn.

I realized that I had seen these trees as an emblem of my own desires, my longing for a natural world that had been starved of healthy fire, but if protected from megafires, could someday return to stability. I had clung to this idea because it gave meaning to my exhaustion, sweat, and pain. Now the forest transformed in my eyes from natural ecology to market economy, then into a wall of flames galloping toward us through the treetops.

※

"YOU THINK THIS IS NORMAL," AXEL TOLD SCHEER AS WE FINISHED THE cigarette. "This isn't normal." He wasn't talking about the bulldozer busy masticating the forest or the spot fire lapping inches from his boots. He was talking about the flames rising from the canopy across the valley. Scheer had complained that this fire season was off to a slow start, but when Axel had joined the Forest Service just a decade ago, a hotshot would consider themself lucky to experience one single fire like this in their entire career. "It's not normal," Axel repeated. "It's just all you've ever known."

Yet everything about the scenario was beginning to feel normal, from the forest that had been clear-cut then cultivated for timber, to the logging boots we wore on our feet. In my pocket, I carried a chainsaw certificate with qualifications drawn by the logging industry to dictate the size of trees I was allowed to cut. Our helmets bore a sticker featuring a pine tree shielded by a line labeled *Incendi*

Proeliatores—Latin for "fire warriors." That line represented us, the hotshots. The extractive spirit of fire suppression was everywhere: lurking in archives, adorning our bodies, filling our language, and stoking the megafire that towered over us through the trees.

Even that fire was beginning to seem normal. Bit by bit, day by day, the megafire felt less like a belching, contorted monster that might swallow us up at any moment and more like a navigable maze whose tricks and turns we could anticipate.

Aoki's voice echoed this normalcy when it crackled on Axel's radio, reporting from his lookout position that we should immediately retreat to the vehicles. "Copy," Axel replied. I expected us to run, but Axel didn't move.

"You know why we're leaving?" Axel asked.

Scheer surveyed the approaching flames calmly. "Yeah." He pointed to the sky. "Looks like that column is bending over us now."

"Bingo." Axel turned to walk back down the road.

I knew that Aoki had a notoriously high tolerance for danger. When he ordered a retreat, the threat was serious. Yet as we shimmied down the hill toward our vehicles, digging the heels of our boots into the dirt, Axel's conversation with Scheer felt like some strange dream. With the megafire racing toward us, Axel and Scheer were debating where to catch the best waves during our next days off. Scheer wanted to book a campsite on the Central Coast, but Axel was planning to go south with his wife and kids.

By the time I jumped into Aoki's truck and he asked me if I had ever been *inside* a forest fire, I felt caught in a schizophrenic dissonance between the extremity of the situation and the normalcy of my colleagues' demeanor. Aoki was as relaxed as he had been in his office the day we met. When the helicopters finally arrived to soak the forest beneath our escape road, Aoki put the truck in gear and drove at a crawl, stopping twice. The first stop was to wait for me to move a burning log out of our path. The second was to take a quick nap while he waited for Axel to catch up in the bulldozer.

With the vehicles safely parked in a torched vale of blackened trees, Axel and Scheer joined us in Aoki's truck. Aoki had an idea. He had found another ridge on his map that looked promising. Maybe we just needed to corral the fire from a different angle. He wanted to check it out.

We descended a maze of dirt roads, bobbing over ruts, then emerged into a sunlit glen on the valley floor. A black bear raced away from us through flowers and reeds. The bear followed the marsh until it reached a meadow lined with bulldozers and firefighters. It ran into the trees. The firefighters were watching the fire.

"There's Cal Fire," Aoki said as we approached, observing the heavyset men with their groomed mustaches and crisp uniforms. "Looking all put together."

Axel snorted. "Hoteled up. Showered."

"Check it out—they're too important to acknowledge us."

"You should get out and slap them with your hair. Get their attention."

Aoki grinned, looking at me and Scheer in the rearview mirror. "Don't mind us. All we really do is drive around and talk shit."

"Check out those Cal Fire dozers," Axel said. Six of them were lined in the shade on the edge of the meadow. "Looks like they've chosen their battleground. Someone should bring them a war drum."

Aoki laughed at this, and we kept laughing as we followed his map away from the fire and the people, down a stream then up again, into a forest where branches screeched against the truck doors. We parked near the top and hiked the rest of the way. Aoki crouched on the peak.

"Besides talking shit, this is all I do," he said, looking over the valley. "I walk around and drive around and look around. It's all about knowing what to see."

From this vantage, we could see the whole of the valley, draped in the golden green of late afternoon. The water was a silver ribbon unfurling into a meadow at the base of the slope we were perched atop. On the other side, the fire was caught in a bowl-shaped depression in the mountains. The smoke billowed in place, glowing in the light.

Aoki seemed pleased with what he saw. The river valley and meadow could reasonably be defended when the fire broke free of the bowl. The slope beneath us was wet with moisture rising from the meadow. If this ridge could be fortified, the head of the fire could be cornered, trapped, and held until it burned itself out. The wind was southwest, the road northwest. Perfect.

Aoki shared this information on his radio. We watched specks of people and machines respond in clusters of movement below. Axel was delighted to see the six Cal Fire bulldozers leave their shade to roll in formation across the meadow. "Look at 'em go!" he called from our peak, where no one could hear. "War drums! Boom, boom, boom, CHARGE!"

Aoki grunted, stalking back to his truck. His plan was in motion. But to know how much time we had to carry it out, he needed to take us inside the wildfire.

WE SAT IN AOKI'S TRUCK IN A GROVE OF BLACK TREES. AHEAD OF US, the road disappeared into a wall of smoke. The smoke obscured the same forest we had escaped several hours before. Strange sounds emanated from within: creaks and crashes and flurries and whines. Within that maw, the fire was alive, and Aoki was looking for it.

Aoki spoke quietly. "Well, boys, we fought the dragon this morning and got our asses kicked. Now we're going to crawl into its mouth to check out its guts."

The laughter of the afternoon faded as we rolled into the darkness, moving through the dead trees that lined the road. Smoke swallowed our headlights. Our vision was reduced to a billow of dust and ash and embers tumbling over us on a hot wind. Swarms of fire beetles pattered against the windshield. I glimpsed a bird on the roadside. It flapped in circles with a burned wing. The images were like snapshots of a dream, swallowed in an instant by shifting clouds. Wind keened through bare branches.

"Apocalyptic," Axel muttered. "Fucking apocalyptic."

Aoki stopped the truck and stepped out, disappearing up the mountain. He was looking for something in the smoke. I was looking for something too. I was looking for sense in the ruin, some pattern in the ashes and embers and death. It was a scene from the darkest corners of my imagination.

As we waited for Aoki, the dragon pressed down on us, groaning, whining, calling. The rote explanations—that these fires were caused solely by hapless government agencies, drought, or stray sparks—obscured the dragon's true form: it emerged from centuries of colonial rule, corporate greed, and political collusion. Fire suppression had, up until now, worked exactly as intended. It had protected the life of profit, at the expense of everything else. When colonialism and genocide opened the forests to logging, the costs of ecological instability were external to most people's lives. We could cut lines around ravaged ecosystems, box them in, hide them, and suppress them so that few people would need to see the results.

But barricading these externalities is the pinnacle of reckless naivety. For decades, centuries even, we could pretend it had worked. Chainsaws and bulldozers could hold the wildfires at bay, keeping them simmering somewhere "out there," far from home. Then something changed: the atmosphere, the temperature, the droughts. Suddenly, the externalities became apparent, spilling over our barricades, burning our homes, our cities, and bleeding into every aspect of our lives. Climate change became not a departure from history, but an extension of it.

This was our legacy as hotshots. Yet the megafires we inherited were already so altered, so powerful, so destructive that they demanded continued suppression. If we let them burn, we would be allowing an intensity of incineration most forests have not evolved to survive. Axel told me this was a paradox we must live with. By fighting fires, we ensured they would continue to grow. But they were already so massive that we had no choice.

Above us, the smoke would be boiling into the sky like a mushroom cloud. Its bulbous head would rise upward until the vapor froze and fell back to earth, flashing with lightning and dropping black rain. Within that beast, the forest had grown dark. When I looked through the window, I could see only a reflection. It was a face. Its cheeks bristled with a beard. Its lips were chapped, skin cut and caked in dirt. Looking out the window, searching for a glimpse of the dragon, I almost did not recognize the image of myself.

※

THAT EVENING, WE FOUND THE REST OF THE CREW IN A GROVE OF burned trees. The hotshots lounged around the buggies in various states of disrepair. Red sat on a stump, staring at nothing. Márlon tended a blister. Barba was asleep, curled in the small space of his world. Drogo loomed over Edgar with a dazed look, listening to him growling a lesson in chainsaw mechanics. Everyone stood and gathered when Aoki arrived.

Aoki had found what he was looking for in the smoke. The fire had thrown embers to the bottom of the mountain, then burned uphill. A downslope wind was pushing it back against its own burn scar, where there was nothing left to feed it. This would buy us enough time to fortify the ridge over the meadow. Maybe we could stop it.

The prospect of more work deflated everyone. Our thoughts were on the days ahead, after the fire. Márlon's fiancée had hired a babysitter and booked a romantic hotel. Drogo was dusting off his suit to visit a vineyard with his girlfriend. Scheer was already surfing in his head. We were almost finished. Soon, we could leave this fire behind.

That night, to the north, winds would blow a tree onto a remote power line. The electrical current ignited the Dixie Fire, which would become the largest megafire ever to burn in California. In less than a month, the Dixie Fire would incinerate this entire forest, swallowing all our victories in a single, unstoppable blaze.

Aoki tried to perk us up. Before we dispersed, he made an announcement: "Good news, fellas. The media is calling us heroes."

Part III

RUIN

CHAPTER 9

Earlier in the season, the Los Padres Hotshots had passed through a desert that was white with noonday heat. The horizon shimmered over flattop mountains. We were on our way home from a lightning fire. Aoki wanted to show us something.

Our caravan turned into a nondescript parking lot off a highway outside Phoenix. We unloaded from the buggies to hike into the hills, leaving our packs and chainsaws behind. We weren't here to fight fire. This was a pilgrimage to Yarnell Hill, the place where in 2013 the entire crew of the Granite Mountain Hotshots, except for the lookout, burned to death.

I first heard of the deaths of the Granite Mountain Hotshots years before, in my early twenties, while hitchhiking with Kenzie across the Southwest. A Diné man named Nez had picked us up just outside Flagstaff and offered to drive us all the way to California, where he was heading to work on a fire. He was employed by the Forest Service to ensure that bulldozers and crews did not destroy

Indigenous archaeological remains. Those sites, he told us—the burial grounds, cliff dwellings, and artifacts—were his history. As a child, Nez had been taken from his mother by Mormon missionaries, who taught him that his ancestry was demonic. As an adult, he had become an archaeologist. Preserving Indigenous artifacts, for Nez, was a way of reclaiming a history that had begun before missionaries, before colonialism, before genocide—a history he hoped could break the cycles of suffering. "A circle is the strongest shape," he told me. "And the most difficult shape to break."

Nez was quiet, with long hair and full cheeks. As we crossed the Great Basin together, he prayed in Diné Bizaad whenever the freeway crossed a stream. He had a dark sense of humor, facetiously translating traditional Diné place-names into names that reflected the new colonialism of corporate governance. "Place of radioactive water," he told me, pointing to a town beneath a uranium mine. "Place of many heart attacks," he added later, passing a town filled with fast-food chains.

Nez grew somber when I asked him about Yarnell Hill. The day the Granite Mountain Hotshots died, Nez was assigned to that same fire. He was supposed to be out there working, but he recalled watching the sun rise red and feeling a preternatural sense of foreboding. The sensation gnawed at him as he made coffee, ate breakfast, and prepared his pack for the day. Nez turned down the assignment and watched smoke whip across the sky. By the time the sun set, the hotshot crew was dead.

My own hotshot crew followed a dirt path for four miles. I heard rabbits scurrying in the brush and vultures flapping overhead, replacing the normal sounds of jokes and laughter. We were heading toward the crosses that marked the sites where their bodies had been found. I suspected that the other hotshots were thinking the same thing as I was: *This could happen to us.*

For hotshots, no death is meaningless. The loss of the Granite

Mountain Hotshots marked a new site on a landscape filled with such tragedies: the Romero Fire, Loop Fire, Thirtymile Fire, Rattlesnake Fire, and many dozens more. Sites of fatalities are magnets for wildland firefighters, who visit to sit in those places, stand in the shoes of the deceased, and reflect on the decisions that resulted in the loss of life. The evening before, at our hotel in Phoenix, Aoki had admonished us not to descend on the bars and clubs of the city, but to remain in our rooms studying the incident reports. We needed to understand the tragedy so that when we arrived at the nineteen crosses, we would know what to see.

Three miles in, when we crested a hill, I noticed that many other crews had made this same journey. Someone had erected a wooden wall that was covered with memorabilia other crews left as tribute: patches from Los Angeles, Tucson, and Tulare, a hat from Tahoe, a shirt from Rock Creek. Edgar told me that firefighters honor the fallen because we gain knowledge from their mistakes. "All of our safety guidelines are written in blood," he said.

Of the Yarnell Hill Fire, this much is known. Just after four in the afternoon, the Granite Mountain Hotshots left their safety zone in an area that had already burned. They attempted to cross through a mile of unburned brush to reach a ranch on the edge of town. Less than half a mile from the ranch, they entered a box canyon. A distant thunderstorm produced outflow winds, pushing the fire up the canyon like smoke through a chimney. The crew was trapped. They had nowhere to go when the fire swept over them.

We descended from the tribute wall into the canyon, following the route that the Granite Mountain Hotshots would have taken. I walked with Red. He was second-in-command under Edgar in Alpha Squad, so our interactions had been restricted by Edgar's strict hierarchy. Now, Red's freckles were burning in the sun, his lip was full of tobacco, and he was talkative.

"The shitty thing is, no one knows what happened here," Red told me. "Normally when someone dies, their crew members can tell you what happened. When everyone dies, no one knows."

I made a sound to tell him I was listening.

"See," he said, pointing to a cluster of buildings at the mouth of the canyon, a quarter mile from the crosses below us, "maybe they thought they could get to the ranch. It looks like you can touch it from here. In reality, there's no way." The brush would have been ten feet high when it burned, all but impossible to cross.

Before I could reply, he carried on. "A lot of people blame the lookout because he wasn't there. But if you read the reports, he had to bugger out too."

I hadn't read all the reports. There were dozens of them, each trying to provide a forensic, step-by-step account of the burnover. It was becoming clear that Red had pored over every page. Axel thought Red was dumb because he wasn't scientifically literate, but I had grown to think of Red as a different kind of smart. His was an intelligence of raw practicality, which gave him a sense of immediacy that is invaluable on the fireline. The Yarnell Hill Fire tormented him because he was searching for answers that didn't exist.

"Other people blame the weather," he continued. "They say it was an unexpected wind shift. But I don't know what was unexpected about it. The weather service *told* them. Everyone copied. Everyone fucking copied!" Red was breathing heavier, though we were hiking downhill. "Everyone wants to blame someone for Sean's death."

"Who's Sean?"

Red turned away, but not before I saw that his eyes were glistening. "My buddy. My neighbor growing up." Throughout his childhood, Red had shared birthday parties with Sean. They learned to ride bicycles together, helped each other in school, and tried out for their high school football team. "Then we joined the Forest Service together. I started on an engine. Sean started on the Granite Mountain Hotshots."

We reached the bottom of the canyon and came to the crosses that marked where the bodies had been found. There were nineteen of them. The crosses were metal, weatherworn, with the names of each person etched on their surface. Red went to sit alone. I met Aoki's eyes and he beckoned me over. He had been here before. Aoki rubbed his skin-grafted hands together and looked at me appraisingly. "What are you thinking about, Thomas?"

A ten-foot-high husk of brush stood, blackened, between the crosses. "I was looking at that skeleton and thinking how much taller the brush was when it burned. How close it was to their deployment site. How hot it would have been."

After listening to Red, I had other thoughts I didn't voice. I imagined Sean disoriented in the sea of green, pushing through brush, branches tearing at his face and clothes until he realized escape was impossible. The moment when his superintendent told him to deploy his emergency shelter. I imagined the final minutes as the fire grew closer, when the crew would have started their chainsaws, cutting brush in a desperate attempt to clear a space large enough to survive.

The crew would have lined up their emergency shelters side by side to reduce the surface area exposed to heat. They would have known from training to keep their feet to the flames, to use their knives to scrape small depressions under their faces to access air that wouldn't scorch their lungs. When firefighters die in wildfires, they usually die of internal injuries, inhaling searing oxygen and burning from the inside out. Some compare the pain to wrapping your lips around a steaming tea kettle and inhaling with deep, panicked breaths.

The Granite Mountain Hotshots would have known to make

noise, any noise, to let the others know they were still conscious, still alive, until everything was silent except for the crackle of embers the fire left in its wake.

◈

BURNING ALIVE IS THE MOST ABSOLUTE VIOLENCE HOTSHOTS FACE, but it is fairly uncommon. Since 2000, of the hundreds of total wildland firefighter fatalities in the United States, only approximately 18 percent of those were attributed to burnovers. Of those, most are cataclysmic events, such as the Yarnell Hill Fire, which claimed nineteen lives all at once. While burnovers, when they occur, produce a sharp emotional response, for firefighters, the acute fear of burning to death is rare. More common is a low-frequency, omnipresent sense of dread that rises from the constant possibility of random catastrophe and the slow violence of financial ruin that follows.

On the fireline, injuries constantly occurred around my crew. At the first megafire of my career, the flames behaved in a way that would have been impossible before climate change, doubling in size overnight. The fire overtook fifteen firefighters on a ridge above me. None died, so they did not add to the fatality statistics. But one firefighter was temporarily blinded by the superheated gases, and another lost the skin on his hands, feet, and face. Doctors amputated several of his fingers. The rest suffered burns.

Other injuries were more common. People frequently collapsed from heatstroke, broke arms or legs, and gashed themselves with chainsaws. A month after the firefighter on our fireline broke his femur, a falling oak tree broke another's spine just after we left the site. The same day, less than a mile away, a woman tripped into an ash pit and lost all the skin up to her elbows. A few weeks later, a cannonball-size boulder hurtled downslope, bouncing inches from Scheer's jaw.

For much of my time as a firefighter, I assumed that I would be taken care of if I was injured in the line of duty. This belief didn't entirely alleviate my sense of dread, but it dulled it. I could break a bone, burn my skin, or collapse from heatstroke, but as long as I didn't die—I thought—everything would work out. I could recover, and I could pay my bills while doing so.

Aoki shattered this illusion on a hot and smoky morning. He gathered our crew outside the buggies for a discussion before we returned to the fire. Two firefighters had suffered career-ending injuries the day prior. Aoki told us their names and encouraged us to contribute to their GoFundMe accounts. He remained stoic, but for him, this was personal. He told us that donations were the only way he had stayed financially afloat after he burned. "Five dollars, twenty dollars, whatever you can spare," he said. "It all makes a difference."

Aoki's statement hit me in my gut. It spread through my body, making my palms sweat. Before we loaded into the buggies, I grabbed Barba's arm. I wanted to know if I had misunderstood Aoki. Were we really on our own? Wouldn't the Forest Service pay for our bills if we were injured?

Barba laughed in my face. "In your dreams, bro." He shrugged me off and climbed into the buggy.

Hiking toward the fire that day, everything looked different. The creak of branches made my eyes dart overhead. Every patter of stones made my stomach turn. Every misstep, I knew, could trip me into ruin. I felt like I had been walking on a tightrope across an abyss, thinking that a safety net would catch me if I fell, only to learn that the net was nothing more than a spider's web.

After this realization settled, I began looking into the aftermath of firefighter injuries and deaths. Earlier in the season, while my crew was fighting a fire in Arizona, a nearby firefighter had died of injuries after parachuting from a plane and landing too hard on a rocky hillside. Reviewing the fatality, I learned that his wife had been charged thirty thousand dollars for her husband's air ambulance flight, three thousand dollars for anesthesia, and over one thousand dollars of miscellaneous bills she didn't understand. She had only been able to stay by his side as he died because of support from a nonprofit, and she was relying on GoFundMe to get by.

The more I looked, the more I found. Dozens of federal firefighters had set up GoFundMe pages to help them recover from torn tendons, broken pelvises, snapped spines, seizures, and heart attacks; to pay for cancer treatment after years of inhaling toxic smoke; and to help families whose main provider had died on the job. One firefighter nearly had to file for bankruptcy after a tree broke his back, costing him two hundred thousand dollars in medical bills. For years, creditors called him eight times a day, then began harassing his family. I searched for the firefighter who lost his skin and fingers on my first megafire and found him on YouTube asking for help. He had been in the hospital for over a month. The Forest Service's HR department told him to set up a payment plan.

How did this happen? One answer lies in federal bureaucracy. Since the early 1970s, hotshots have been classified not as firefighters but as forestry technicians, which is considered seasonal contract work without benefits. The United States Forest Service created this job classification to establish a flexible seasonal workforce that could meet the diverse needs of ecosystem management—maintaining trails, restoring watersheds, and occasionally suppressing fires when needed. When I spoke with personnel from that era, they recalled summers when they wouldn't be called to fight a fire until August. Fire suppression only consumed approximately 10 percent of their time.

In the decades since, carbon emissions have packed millions of years of environmental change into a single human generation, and the institutional classification of Forest Service firefighters has not kept pace. In the early 1970s, when the forestry technician classification was created, atmospheric carbon concentrations hovered around 320 parts per million—at the upper end of historical averages. By 2021, when I was a hotshot, those carbon concentrations had risen to 420 parts per million, higher than at any point in human history. Since 1988, when scientists told Congress that fossil fuels would drive catastrophic changes in our climate, more carbon has been burned than the cumulative emissions from ten thousand years of human civilization. This point bears restating. The fossil fuel industry has produced more carbon in three short decades than was emitted in all of human history combined, with knowledge of the consequences.

The growth of wildfires, and the work demands of hotshots, tracks the rise in carbon emissions. During the 1990s, only 16 percent of the Forest Service budget went to fire suppression. In that climate, it made sense to classify employees as forestry technicians because they fought fire for only a small fraction of their time. But in the following years, wildfires have grown exponentially. By 2021, fire suppression consumed around 70 percent of the Forest Service budget, and 90 percent of hotshots' time. Megafires that before climate change would have been once-in-a-career events now burn regularly across the American West, for months on end. Federal firefighters, like many frontline communities, are trapped in a paradigm of a different planet. They are saddled with the work expectations and low compensation of a bygone geological era, an era when fires started later, rains came earlier, and record heat waves were centennial rather than annual occurrences—an era that ended when the fossil fuel industry pushed our planet into uncharted territory.

Because federal firefighters like hotshots remain classified as forestry technicians, their base pay starts at fifteen dollars per hour. Most hotshots' compensation only exceeds minimum wage when they receive hazard and overtime pay. A "perfect paycheck" is 112 hours of firefighting in a week, or sixteen hours per day, which is the maximum allowed. (Federal employees, unlike state and county firefighters, are not paid during their eight hours of sleep, even when they are sleeping on the ground in thick smoke.) To pay their bills, hotshots need hazard and overtime pay, which means they need to spend as much time as possible fighting increasingly dangerous fires. Yet when they are injured on these fires, they lose the hazard pay and overtime pay that provides them with a livable wage. During recovery, they receive approximately 80 percent of their base pay. For a beginner like me, that would amount to twelve dollars an hour. Even for a hotshot superintendent like Aoki, the loss of income can be catastrophic.

However, federal bureaucracy only partially explains hotshots' deteriorating working conditions. Also to blame are the vested interests that deny climate science, delay climate action, and erode the rights of people dealing with climate change's deadly consequences. Koch Industries is a case in point. As the second-largest private company in the United States, Koch Industries captures annual revenue larger than that of Facebook, Goldman Sachs, and U.S. Steel combined. Much of this revenue is derived from fossil fuels. Charles Koch, the CEO and the twenty-fifth richest person on earth, spent over $145 million attacking climate science and policy solutions between 1997 and 2018. The spending was effective. In those same years, climate change denial spread among the U.S. public, the federal government failed to pass any substantial climate legislation, and fossil fuel corporations claimed record profits.

Even as corporations' rampant climate change denial makes working conditions more dangerous, these same corporations endeavor to eliminate workers' rights. In the past several decades, the American Legislative Exchange Council—a Koch-funded group that drafts state legislation—has passed policies that ban local governments from setting their own wage laws, ban cities from setting decarbonization goals, cripple collective bargaining power, and fight guaranteed access to health care. Similar corporate groups also fund the elections of officials who adhere to the mantra that the role of the government should be to get out of the way of corporate interests. Grover Norquist, a Koch affiliate, put the goal bluntly: "I don't want to abolish the government. I simply want to reduce it to the size where I can drag it into the bathroom and drown it in the bathtub."

The result on the ground is that the federal employees charged with containing disasters caused by corporate malfeasance are drowning in divestment campaigns funded by corporate malefactors. In Congress, reclassifying hotshots as official firefighters has proven impossible, because it requires a bipartisan coalition to vote to use tax dollars to provide hotshots with health care and living wages. Some in Congress do not feel that hotshots' work deserves better compensation. As California's Republican congressman Tom McClintock said, "Wildfire firefighting is hot, miserable work, but it is not skilled labor."

Soon after McClintock's statement, I sat alongside Drogo in a helicopter. We were circling a plume of smoke in McClintock's district. The helicopter dropped us off in a meadow near the fire. Our task was not to put the fire out, but to use MIST—Minimum Impact Suppression Techniques—to steer it away from a reservoir that nearly three million people depended upon for water. These people included Congressman McClintock's constituents. Using hand tools and knowledge of the landscape, we worked alongside the flames, digging trenches between aspen groves and rocky outcroppings. Our fuel break kept the fire in areas that benefited the forest, and we eventually steered it onto a granite ridge where it was naturally extinguished. McClintock's reservoir remained unscathed.

That night, as we huddled around a campfire in the forested fringe of a meadow, I asked Drogo what he thought of McClintock's stance. Drogo just shook his head, staring into the flames, face covered in soot.

McClintock may be only one lawmaker, but he represents the insidious bind of a political process funded by the fossil fuel industry. McClintock denies basic climate science, calling it "suspect." He has signed a pledge to Charles Koch to functionally oppose any legislation that would combat climate change. Simultaneously, he blocks funding from the very workers whose lives are imperiled by his climate change denial—even as those workers protect his constituents' watershed.

As we traversed the state, we often drove beneath overpasses where residents had hung banners thanking us for our work. At the entrance to many small towns, hand-painted signs called us heroes. On one fire, after we saved a mansion worth four million dollars, the owner walked over to us and tried handing each of us a $100 bill. This facade of individual gratitude is particularly cruel when it overshadows the systemic reforms necessary for protecting firefighters' lives. One of my crew members told the owner of the mansion to keep his dollar bills, but to please keep paying his taxes.

"Banging on cans and cheering firefighters as heroes is a poor substitute for what they should clearly earn," writes James Puerini, a hotshot veteran and forestry researcher at Yale. "Give heroes what they deserve—affordable healthcare."

UNDER THE HEAT OF THE ARIZONA SUN, SEATED AMONG THE CROSSES of the Granite Mountain Hotshots, I thought the idea of sacrifice was apt. But while Edgar explained that hotshots' lost lives were a sacrifice that allowed us to learn from their mistakes, my mind was following a different path. We were sacrificing the safety of our bodies to rein in conflagrations produced by the fossil fuel industry, and we were sacrificing our rights to fair compensation so that the costs of taxation wouldn't decrease corporate profits.

The precarious lives of hotshots are one flashpoint in an expanding field of self-reinforcing social and environmental crises. Scientists call this a sacrifice zone—a place where low-income people shoulder the burden of industrial misconduct. For hotshots, just as burning to death is less common than catastrophic injury, catastrophic injury is less common than the chronic health effects of the work. Because taking time off to seek medical treatment would mean losing the hazard and overtime pay hotshots rely upon, many tend to their own medical needs covertly. One of my crew members worked for years with a torn MCL ligament in his knee. I knew another hotshot who worked with a hip joint that was eroded to the bone. He could hear his femur grinding against his pelvis as he hiked. He visited a veterinarian to get painkillers for his dog, then took those pills to treat his own pain. The last time I saw him, he was pissing blood, but he was still fighting fire.

"Sacrifice zone" usually refers to a geographical area with distinct boundaries, like a town near a chemical plant. Instead, I began to think of sacrifice zones as concentrated in our bodies. Like the particles of smoke that clung to our hair and clogged our lungs, we carried them with us wherever we went. Cancer is the leading cause of death among wildland firefighters. Each year, the U.S. Department of Labor receives approximately 2,600 workers' compensation claims from federal firefighters, about 175 of which are for cancer and heart and lung diseases. The real damage is likely much higher. Until 2022, for federal firefighters to have received support for chronic disease, they must have definitively proven a link between exposure to carcinogens and their particular ailment. This meant filing paperwork each time they experienced a hazard, even as they spent their days and nights working and sleeping in clouds of smoke. Every time we hiked away from a wildfire, the fire was inside of us, in our bloodstream and lungs.

Just as sacrifice zones suffuse our bodies, they also affect those we love. Later, during a period of rest, I went to a barbecue at Scheer's trailer, tucked in an avocado orchard in the mountains. I approached the girlfriend of a hotshot I knew who had joined a different crew. I asked her if she was okay. Her whole face collapsed. She told me she had been numb for the first few weeks, but late one night, his absence crashed down on her all at once. She sobbed for hours in bed, alone.

Throughout the fire season, Jack was one of my favorite presences on the crew. At first, no matter how hard things got, you could always count on him to make you laugh. Yet as the months passed, he became more reserved, as if a weight were drawing him inward. Midway through our fire season, his "old lady" left him during our mandatory three days of rest. She hadn't looked up from her phone during dinner one night, Jack had accused her of being ungrateful, and the situation had devolved from there. When she left, she took their toddler daughter. Between assignments, on Jack's first night home, his daughter would straddle his chest like a koala bear and ask if he was going to fight fire tomorrow. "No, sweetie." The next night, she would ask again. "No, sweetie." On the third night, he'd have to tell her, "Yes, sweetie, daddy's gotta work tomorrow." She would start crying and clench herself to him until she fell asleep. Now his daughter was gone. He suspected his ex-girlfriend was with another man. Such domestic tumult is common in the world of hotshots, who claim, only half jokingly, a 300 percent divorce rate.

Nelda St. Clair, a firefighting veteran and the wife of a hotshot superintendent, surveyed wildland firefighters to try to better capture these statistics. She found that almost half the partners of wildland firefighters considered ending their relationship because of conflicts that stemmed from the job. Almost three-quarters of those partners were anxious that their loved one would be injured or killed fighting fire, while only 11 percent felt confident that the Forest Service would help them should such a tragedy occur. Over half reported feeling secondary to the job. "[That was] the most difficult part," reported one partner. "We always came second. Fire always came first."

While home, according to their partners, only 37 percent of wildland firefighters seemed at ease. Less than half could sleep through the night, and most suffered mental health challenges from their job. And climate change was making everything worse. "This season is the fastest I've seen him burn out mentally and emotionally," one partner reported.

As this survey was being conducted, I spent an afternoon on the fireline with Scheer. We were waiting for Edgar and Márlon to extinguish a persistent patch of flames in a gully below. We sat at the mouth of a cave, watching the shadows of the forest turn with the arc of the sun. Scheer was using my phone to play a Charley Crockett album on repeat, passing the hours with twangy blues.

I found myself in an emotional limbo: while I was on assignment, I wanted to be home, but while I was home, I couldn't wait to get back to the crew. At home, my attempts to connect with Kenzie felt increasingly hollow. The last time she had dropped me off at the station, for this assignment, we hadn't kissed. We hadn't even said goodbye. Our days together had been labor. She had planned long walks and dinner dates, but I had greeted her with a hacking cough, dark mood, and fragmented stories. During one story, I noticed she was crying. At the end of those days, when she dropped me off at the hotshot station, I accidentally slammed the car door. I cracked it open to apologize, then accidentally slammed it again. I didn't look back until I had passed through the hotshot gates. Kenzie had already driven away.

My thoughts often wandered in these directions during those long afternoons. I wasn't the only one. At the mouth of that cave, I glanced at Scheer. I was shocked to find him seated, holding his knees, with tears drawing lines through the dirt on his cheeks. I asked him what was wrong. He tried to laugh, but his voice was garbled. He fumbled for his phone for a moment to pull up a note, then explained himself with a poem he had written:

I sat on the mountain and thought of days gone past
Wondering how long all this pain is gonna last
Will my soul still be on that dusty road
Out West with my brothers watching ourselves grow old
I'll have to keep panning hard till I find that gold
I smell the pinon burn and see the fire rise
All the way up until it hits that desert sky
As day fades to night and the sky comes alive
I'm still dreaming of that girl and I don't know why.

He was in love. He was worried about his new girlfriend. He was tired. He had given a substantial amount of his money to family members to help them pay their bills. I wondered if the burned bodies of the children he had encountered two years ago were somewhere in his troubled mind, dissolving in that mix of emotions.

I spoke with other wildland firefighters who had discovered burned bodies in the forests. One found the remains of a man outside an incinerated house. Another recalled walking through the ruins of the town of Paradise after it burned, seeing charred bodies mixed in the rubble. "We're not trained for that," he told me. "We're trained to cut trees and backburn and dig line, not recover corpses."

Every year, at least as many wildland firefighters die of suicide as on the fireline. Just as wildfires can't be pinned to a single cause, firefighters' suicides often result from multiple factors that accumulate to a breaking point: increasing work demands, economic neglect, strained relationships, and intermittent trauma. For their loved ones, these expanding zones of sacrifice are real. "If a fire doesn't kill my husband," said one hotshot's spouse, "the stress and anxiety will."

"Who do you think will do it?" Drogo asked me one day, across the aisle from me in the buggy, without meeting my eye.

"Do what?" I asked.

"You know."

For Drogo, as for many hotshots, the idea that someone on our crew would resort to self-harm seemed an intuitive threat. The relentless accumulation of pressures was pushing the limits of my own capacity to cope.

I texted Kenzie a message I had never sent before: *I'm scared.*

CHAPTER 10

THE SOUND OF RUSTLING BODIES WOKE ME ON THE SHORES OF A swamp. We were past the midpoint of the season, so the first sleeping body Aoki kicked usually roused everyone like an alarm bell. But that morning, we rose slowly. Even those of us who could sleep through the cockroaches scurrying over our bodies had been sapped of our energy by the nighttime humidity. My eyes were crusted and my tongue was thick, like from a bad hangover. My shirt was still soaked with the sweat of yesterday's work. As I pulled it on, I noticed a piece of flesh missing from my wrist, like someone had burned me with a cigarette. Two red pinpricks marked the middle. It looked like a spider bite.

According to protocol, we were supposed to report any physical abnormalities up the chain of command before they became a problem. I started with Jack, who was making animal noises as he pulled on his boots. He blinked at me through his tangled hair and grunted. I took his grunt as confirmation of my spider diagnosis, so I moved on to Márlon, who told me through a mouthful of toothpaste to "put some saliva on it." After spitting on my spider bite, I approached Red. He was only two years my senior, but he groaned like an old man as he stood from his sleeping pad beside the buggy and took my wrist in his hands.

I was self-conscious. Reporting something as simple as a spider bite went against everything in my upbringing, but Red's eyebrows were furrowed with genuine concern. He told me to circle the rotten flesh with a Sharpie. If the skin kept retreating, that would mean the bite was from a black widow or brown recluse, and he'd get me help. I'd be sticking with him today, just to be safe. Red shuffled off to report my situation to Edgar. When I brought Edgar coffee, he adopted a baby voice and asked me if I had a boo-boo. But his lips curved up at the corners in a semblance of a smile. Edgar liked me more now that I was a puller.

We were in for a boring day. As Northern California descended into something like a climate apocalypse, with the Dixie Fire pushing a million acres and two more megafires erupting around it, we were stuck in a desolate corner of Kern County babysitting a small fire that was already contained. Approaching the fire, we drove over barren earth through a chemical haze. Thousands of shapes moved in that haze: oil derricks, pumping up and down like seesaws. Many of those derricks were serviced by Halliburton, a corporation that Dick Cheney presided over as CEO before he became vice president of the United States. Halliburton had sucked billions of dollars in federal contracts from the war in Iraq. As our buggies left the oil fields for the foothills, those derricks left an imprint in my mind. They were out of sight but always there, moving the underground to the overground, contributing to the blanket of heat that pulled water from the forests and fire from the land. I rubbed the open sore on my wrist, looking out at a forest canopy more brown than green.

In the back of the buggy, I joined Scheer and Drogo in a huddle. They were talking about a more immediate conspiracy than that of fossil fuel corruption. I turned on some loud Nicaraguan music, which earned me a thumbs-up and a big grin from Márlon, and also ensured that Edgar couldn't hear our gossip. South Ops was hoarding us, Scheer speculated, keeping us on bullshit fires so we would be available if one blew up in Southern California. South Ops, the Southern Operations Command Center, was a group of people in an office building in Riverside, California, who handled the logistics of the region's wildfire resource distribution. Our crew was one such resource, along with the bulldozers, airplanes, and matériel of fire suppression.

Hotshots think of the Ops as omnipotent deities. The Ops have the power to hook us up or screw us over. They could ship us to Alaska or flip us to Montana. They could plow us into megafires, or they could keep us sitting on small ones for weeks as a rear guard while everyone else fought the monsters in the north. North Ops had a reputation as a generous god. The Tahoe and Sierra hotshots always got exciting assignments in exotic regions. South Ops was more conservative. Scheer said they wanted to hoard us for initial attacks. Since we were among the most skilled hotshot crews in the United States, the Ops needed us to be available to put out small fires so they wouldn't turn into big ones. California couldn't handle another big one. Not now.

Logistically, this made sense. By August 2021, both federal and state resources were stretched to the breaking point. California was spending over six million dollars of taxpayer money each day fighting to contain the Dixie Fire. Meanwhile, federal resources were so emaciated that officials were forced to make risky decisions. They decided to let one lightning fire in the Eastern Sierra burn itself out, not just because it was remote and surrounded by natural barriers, but also, in part, because they didn't have the money or personnel to contain it. This strategy could have worked in the stable climate of past generations, but record heat, brittle forests, and anomalous winds pushed that fire over a granite ridge. It took off, growing into yet another large fire—the Tamarack Fire—and drawing the ire of Cal Fire and politicians alike. Tom McClintock, the congressman from California who had described wildland firefighting as unskilled labor, delivered a speech to Congress, telling legislators that "fire is not our friend." For the rest of the year, it was war. Every fire would be fought until it was dead out.

From the perspective of the Los Padres Hotshots, this campaign of total fire suppression should have been great news. A monster fire like the Dixie would guarantee steady work for at least a month. But our unexpected task of chasing small fires was more precarious. We *needed* fires to keep igniting so we could keep paying our bills. In the back of the buggy, in a whisper that somehow still boomed, Drogo said he doubted this fire would be hot long enough to get us through a whole two-week roll. There were some ash pits still smoking and some big trees holding fire in their trunks, but nothing we couldn't take care of in a few days.

This was bad. This meant that South Ops could send us back to our station before we had worked enough time to earn three days of guaranteed rest. If we returned to our station before our two weeks were up, our schedules would reset the moment our boots hit home turf, and the fourteen-day clock would start ticking all over again. If we kept working small fires, putting them out quickly, we could go months without any guaranteed days off. Even if we were lucky enough to have a day at home between shorter assignments, we couldn't plan dates, couldn't unwind, couldn't get drunk, because we had to stay "fire ready." Scheer couldn't even surf, for fear he would miss Edgar's dreaded text: *Wheels rolling in two hours.*

We parked on the shoulder of a road and trudged up the fireline. On one side of the line, a blanket of ash covered the ground under the trees. The other side was a jungle of shrubs. Red and I dropped off toward the bottom, probing the ashes with the backs of our hands to look for heat. Everything was cold, so Red settled against a log for a long afternoon. I sat beside him. We could hear Edgar's chainsaw buzzing from deeper in the forest, with the *creak, crack,* and *boom* of falling trees.

Red was a naturally easy presence, but it was rare to see him relax. In the months he had apprenticed under Edgar, Edgar had carved out a space in Red's mind, like a mental watchtower built of wraparound sunglasses and a square jaw. Red's beard was growing in. He had lost weight. He looked good. I told him so. He said thanks. His friends asked him what his secret was. He told them it was easy. All they had to do was put on a pack and walk up a mountain every day.

Red checked my spider bite once every hour until he decided it was nothing to worry about. I asked him why we weren't assigned to the Dixie Fire. Red despised gossip, so he brushed off the question. "Dunno. The Dixie already has five thousand people fighting it," he said. "They're probably just keeping us fresh for the next round."

I was bored, so I kept prodding. "Some people say Aoki is keeping us away from it because he doesn't like working with Cal Fire." *Some people* obviously meant Scheer, who told me Aoki had friends in South Ops and knew how to pull strings for the assignments he wanted.

Red sighed, resigning himself to my pestering. "Yep. There's that too. A lot of the time, there's trouble working in unified command." Wildfires enter "unified command" when they burn across jurisdictions, requiring state and federal agencies to fight them together. "People butt heads, don't see eye to eye. That's how it used to be, for sure, but there are still remnants."

I had heard stories. The culture clash between federal and state firefighters trickled down from the command tents to the fireline. A Cal Fire engine guy had told me he'd watched a federal hotshot crew cut line, then leave the line to chase a spot fire. A state crew had stepped in to finish the job. When the federal crew returned, they saw this as an insult and spent the rest of the afternoon fixing the work of the state crew, which they said was too sloppy. "It's like having two rival teams in different tents planning the same job, without talking to each other except to yell insults at each other," the Cal Fire guy had said.

From Red's perspective, a lot of the conflict was cultural and came from having different areas of expertise. "We have different ideas about fighting fire," he told me. "Cal Fire may tell us to backburn when we could still get it direct. Or they may tell us to go direct when we know the fire behavior is too extreme and we should be burning. It's a mess."

"Sounds like it."

"To make it worse, a lot of the time Cal Fire thinks us feds are just letting things burn or making fires bigger to get paid more. But the thing is, Cal Fire doesn't really know the mountains. They specialize in the interface."

Red was talking about the WUI, which stands for "wildland urban interface." This is where, he told me, dumbass people build homes in high-risk places. The WUI was Cal Fire's specialty. Red reiterated that they didn't really know the mountains.

"Not like us," I replied.

"Yeah. And they don't work hard. I mean, you saw those guys panting up the line the other day. Jesus Christ, it was hardly a slope!"

I laughed.

"So yeah, might be something to that rumor. Aoki doesn't like dealing with hostilities. The old red [Cal Fire] versus green [Forest Service] rivalries. Not to mention it's a clusterfuck. Cal Fire takes all the hotels. Contractors fill up the camps. Hard to even find dirt to sleep on with so many people."

Red's voice had slipped to a murmur. He was drifting to sleep, so

I moved to a nearby boulder to read a *New Yorker* novella on my phone. While I perused my screen, a Snapchat popped up. A friend from my first year on the beginner crew was on a federal engine now and had captured a video of the Dixie Fire at night, flames reaching for the stars.

"How is it?" I typed.

"Epic," he wrote. Three dots showed he was still replying. "No one here has ever seen anything like this. How's LP?"

Red was snoring. The day was dragging. The most exciting event was when Scheer marched into the forest on the green side of the fireline to take a shit and found a smoldering patch of leaf litter. A candle-size flame was licking inches from brush that would lift flames into the trees. Scheer scuffed it out with his boot. We were one shit away from having a crown fire erupt behind us. I was surprised by my disappointment.

Red's voice jolted me from my thoughts. "Thomas, c'mere. Quick!" His country accent had come out. He was panicking. I hefted my hoe, excited, ready to fight fire. "No! Leave yer tool! Hurry!" Confused, I found Red lying on his side, panting, helmet thrown away, balding scalp gleaming with sweat.

"Goddamn bug crawled in my ear!" He handed me a canteen. "Quick! Drown that sucker. I can hear it scratchin' on my eardrum."

I took the canteen. The folds of his ear were full of dirt. "You sure?"

"Yeah! C'mon! Quick!"

I poured in a thimbleful. Red snorted and shook his head then paused, listening. "Yep! Still in there scratchin'! C'mon, Thomas. Drown it this time!"

I poured the whole canteen into Red's ear. He blustered through the water, then stood, head cocked, listening. "Still in there. I'll get you, you bastard. Hang tight here, Thomas." Red hefted his pack and ran up the fireline to find Barba, who probably, for some reason, had tweezers.

I checked my Snapchat again. "How's being a hotshot?" my friend on the Dixie Fire repeated.

"Epic," I replied, putting away my phone.

WE WERE IN OUR BUGGY WHEN WE HEARD A VOICE ON THE RADIO AN-nounce that we were dismissed from the fire. We had put it out, so we were no longer needed and were to return to our station. It was the eleventh day of our two-week assignment. We needed to work for only three more days to earn our guaranteed days off. This felt like a particularly cruel form of torture, to be so close to the luxuries of home only to have them dashed in an instant by a squawking radio.

The reactions of the hotshots varied. Drogo got my attention and leaned across the aisle, telling me to observe the textbook stages of grief. "Denial," he said, nodding toward a younger hotshot who was babbling that maybe we'd get sent to another fire. "Anger," he said as Smitty punched the back of his seat. "Depression" brought my attention to Scheer, who was clutching his head in his hands and cursing South Ops under his breath. "Acceptance" was Márlon, who hadn't reacted at all, but simply continued watching a movie on his phone. Drogo jabbed two thumbs into his own broad chest. "Hope," he said with a grin. He produced his phone, called his girlfriend, and told her to drive toward us, light cigarettes, and flick them out the window.

It was a peculiar sensation, lamenting a lack of fires to fight. The emotions were inherently contradictory. I never spoke with a firefighter who hoped for devastation, but everyone hoped for a "good season," which meant constant fires. We needed to fight fires to earn a living wage and to get days off, which kept us sane. One engine guy with a glazed, thousand-yard stare once told me about his dream of a firefighter utopia. "A never-ending initial attack," he mused, staring into mountains that would soon burn in the Dixie Fire. "Never-ending, all year, year after year, forever." To me, it sounded a lot like apocalypse.

As deranged as the engine guy's firefighter utopia may sound, it hews closely to the ideals of some of America's foremost philosophers. In 1906, just after the Forest Service was founded, the pacifist philosopher William James stood at a lectern in Stanford. He looked like many philosophers from that time, with tufts of gray hair sprouting like bat wings from his temples and a beard to match. But he was delivering a unique message. In a voice that was clipped and precise, James advocated for a shift in warfare. War, in his view, was an intrinsic part of humanity, so humans should wage that war on nature instead of one another. "Instead of a military conscription," he explained, America could begin "a conscription of the whole youthful population, to form for a certain number of years a part of the army enlisted against Nature."

A war on nature, James elaborated, would produce manifest benefits. The young men of the day were too soft. The war on nature would toughen them up. They would learn higher ideals. Discipline. Hardiness. "A martial type of character." Women would adore them. By taking the fight to the woods, young men would have "paid their blood tax," James predicted, and "done their own part in the immemorial human warfare against nature."

James should have been careful what he wished for. He died before the world wars, so he could scarcely have imagined the destruction that humans would unleash against one another, nor the ramifications of turning this destruction against the land. By the end of World War II, after dropping two atomic bombs on Japan, Harry Truman proposed a "ceaseless war upon the fire menace." The *Los Angeles Times* called for "machines of war to be turned against fire." Paratroopers became smokejumpers. Military helicopters were conscripted to drop water. Airplanes that had dropped bombs on cities were retrofitted with fire retardant. By 1961, when departing president Dwight D. Eisenhower warned that the profitability of the American military would create a self-perpetuating monster that fed on the same conflicts it produced, James's proposed war on nature had become a reality. But instead of supplanting the military-industrial complex, as James had hoped, the war on nature grew in tandem with it.

One man at our station had lived through the growth of the industry of fire suppression. His name was Mark Linane. Though he was mostly retired, everyone called him Supe because he had formerly spent almost thirty years as the superintendent of the Los Padres Hotshots (the position now held by Aoki). Linane was in his eighties. He still kept a water tank and fire hose in the bed of his pickup truck so he could mount initial attacks. That year, however, there weren't many fires near Santa Barbara, so he had to content himself with pestering Aoki. Linane must have been monitoring the radio traffic from his home, because he was waiting for us at our station when we came back from fighting the small fires.

Linane walked with the stooped, shuffling walk of my grandfather, who was also an octogenarian too stubborn to use a cane. A mustache had hidden Linane's upper lip since he was twenty years old, brown turning to white. I never saw him without his hotshot ball cap. During our training, Linane had sat in the back of our classroom, *harrumph*ing at any mention of fire models or other modern technologies. Nothing, he growled, could replace experience. Linane could tell you every time a fire model had failed where hotshot knowledge could have prevailed. He could also show you the scars on the trees from a fire that had nearly burned down our station decades ago. The only thing that saved the station, he claimed, was courage. That, and the rat piss in the walls.

When Linane became the superintendent of the Los Padres Hotshots in 1973, the corporations that profited from the Vietnam War were shifting their focus to America's public lands. The arms industry began selling explosives to the Forest Service. When helicopters transported Linane's crew to a fire, instead of dropping the hotshots in a meadow, they would bomb a hilltop and land in the crater. Meanwhile, Monsanto, among the world's largest petrochemical corporations, began selling off its excess cache of Agent Orange to the U.S. Forest Service in the guise of fire prevention. The carcinogenic chemical was an herbicide; it could annihilate vegetation, creating fuel breaks. Military veterans who joined Linane's crew had the peculiar experience of watching familiar airplanes shower California's mountains with the same biochemical weapons they had used to root out Viet Cong fighters.

In the decades since, America's wildland firefighting effort has further imposed the machines, organizational structures, and mentalities of war onto the nation's forests. Wildland firefighters use a military command-and-control system, with incident commanders in base camp and hordes of subordinates in the mountains. We are equipped with Black Hawk helicopters, drones, incendiary grenades, and flare guns. Even our language mimics war: "initial attack," "extended attack," "search and destroy." To keep track of pay, hotshots use pocket calendars that begin with a short propagandistic history that mistakenly locates the origins of fire suppression in World War II, when firefighters joined a national security effort to prevent the Japanese from firebombing western forests. We are sustained by MREs (Meals Ready to Eat), suspiciously unspoilable calories arriving in glossy brown bags from the U.S. military, where they had been repurposed from the so-called War on Terror. The label reads: "Warfighter Recommended, Warfighter Tested, Warfighter Approved."

When perpetual conflict means perpetual profit, the goals of war blur. For hotshots, wildfires mean they can pay their bills. But for a few industrialists, the bigger the fires get, the more money they make.

WE WEREN'T AT THE STATION FOR LONG. SOUTH OPS SENT US FOURTEEN hours north, where we listened to meth labs explode in a wildfire outside Redding, waited for COVID tests in a grim parking lot, then drove fourteen hours back home because half our crew had contracted the virus. By the time we rolled into a fire camp in San Bernardino another five hours south of Santa Barbara, we were a skeleton crew. Axel called us the V Squad, because only the vaccinated hotshots were left. He joked that the average IQ of the LP Hotshots had just doubled.

The fire camp was in a public park at the base of the mountains outside Los Angeles. The sun had set. Crickets were chirping in the oak trees. Drogo and I hopped out of the buggy and walked down the camp's main avenue to search for crates of Gatorade and bottled water to restock Aoki's truck. The camp had a carnival atmosphere—not the kind with colors and music and dancing, but the dirty, seedy kind of carnival that appears out of nowhere in small towns on humid summer nights. The floodlights. The trailers. The hum of generators, the rotten sweet smell of porta-potties, and the wrinkled men smoking cigarettes in the shadows.

Fire camps are organized based on the Halliburton model from America's wars in the Middle East. Private corporations specializing in disaster response contract with the U.S. Forest Service to create pop-up cities that serve as a home base for firefighters. An army of businesses occupy fleets of trailers. One trailer serves meals, usually in the form of gummy, nondescript meat coated with congealed gravy. Another trailer does laundry. A medical trailer delivers antihistamine shots to an assembly line of poison-oak-afflicted buttocks. Snores come from an air-conditioned semitrailer stuffed with bunk beds. From another emanates the *beep*s, *click*s, and *whir*s of industrial printing machines, which can spit out ten thousand pages a day for distribution to the different crews. Private security officers patrol yurts where commanders handle logistics, financing, and media relations. Another yurt houses the private specialists who create the fire maps. A tent city spreads around the perimeter like a field of rainbow-colored hillocks. For many firefighters, these camps are the biggest cities they have ever slept in.

I stood with Drogo in line outside a refrigerated trailer that we suspected held the Gatorade. Drogo loomed over the two men in front of us. He studied them. They wore fire-resistant long-sleeve shirts, standard yellow, starburst-colored, without a speck of dirt, untucked and unbuttoned. Underneath, they wore T-shirts that a vendor had been hawking at a fire camp in the north. *Tennant Fire*, the shirt screamed in cursive font, beneath a flaming skull. Our hotshot crew had contained that fire.

"You boys ready for night ops?" Drogo asked, nodding at their headlamps. The question was sarcastic. Every wildland firefighter knew that your headlamp belonged in your pack unless you were cutting line after dark. You only wore your yellows on the fireline, and they only stayed yellow if you didn't sweat enough to turn them brown. Even worse, no one worth their salt would ever buy a T-shirt from a fire camp. In spite of these cultural infractions, Drogo's voice was deceptively friendly. He enjoyed lulling people into a sense of ease.

The men took the bait, staring up at Drogo. No night ops for them. They were on a contract crew. They had been fighting fire all day.

"Wow." Drogo nodded with mock interest. "You guys been slaying dragons all summer, huh?"

They had. Like most private firefighting contractors, their company was from Oregon, but they had been in California for months. They didn't like all the COVID restrictions in California, but they liked the fires. Business was good.

Drogo stopped nodding, but his eyes stayed wide as he pivoted away from them, facing me. "Kooks," he rasped in a voice just loud enough for them to hear.

Since the 1980s, both the U.S. military and the Forest Service have transitioned to a privatized system of warfare. In 2020, of the military's $778 billion budget, over half went to private corporations—Halliburton to manage the camps and private contractors like Blackwater to handle the conflict. Profits from the war on nature run parallel. In 2021, federal fire suppression cost taxpayers about $4.4 billion, around half of which was captured by industry.

As disasters have grown and public agencies have been gutted, corporations have rushed in to gain government contracts, funded by taxpayers, to fill the gaps. Private corporations often provide approximately 40 percent of wildland firefighting efforts in the United States, with over 150 firefighting companies employing some 12,000 crew members. The business model is lucrative. As contractors, firefighting companies can bypass already scant labor laws, providing low hourly wages without medical or retirement benefits. The money saved from dodging regulations should, theoretically, lower fire suppression costs. In reality, the money pads profit margins and is allocated to business expansion. In 2022, the Forest Service awarded $640 million in federal contracts to private firefighting crews in Oregon alone. Grayback Forestry, among America's largest wildland firefighting corporations, captured $180 million of those allocations.

While profits flow in from both fire camps and private firefighting businesses, the largest profits are earned in the sky. The next morning, as I huffed up a mountain that was "nuked"—everything incinerated down to the soil—with Scheer in front of me and Drogo on my heels, there were no contract crews in sight. We were too close to the fire. But the air roared with other contracts—giant DC-10 commercial airliners careening through valleys, Chinook and Black Hawk helicopters thumping overhead, and one renegade single-engine crop duster buzzing so close that my teeth rattled and Axel dove to the ground.

The airplanes spew Phos-Chek, a bright red fire retardant developed by Monsanto and owned by Perimeter Solutions, a billion-dollar corporation with a monopoly on the fire chemical market. In 2021, the U.S. Forest Service purchased fifty million gallons of Phos-Chek at around $2.50 per gallon. Over a third of that retardant was dumped on the Dixie Fire, where it was least effective. Phos-Chek doesn't put fires out. It can slow them down, given the right conditions. On low-intensity fires, retardant can give hotshot crews enough time to cut line around the fire's edge. On high-intensity conflagrations like the Dixie, the flames cross the retardant like a tsunami over a seawall.

Fire retardant, when used on megafires, often serves a different purpose. For a public conditioned to viewing wildfire as war, the sight of airplanes zooming overhead to bomb the enemy signals that public officials are taking the threat seriously. Firefighters call these demonstrations of force "election air shows" and "publicity stunts." A forest ranger defended the practice, telling me that he faced tremendous political pressure to call in airplanes during fires, even when they weren't useful. "That's what the public has come to expect," he said. "During the fire, everyone kept saying, 'Where's the retardant, where's the retardant?' And I said, 'Well, just get five air tankers and put 'em in formation and fly 'em low and slow to show 'em we're here.'"

It's an expensive air show—"more expensive than dropping Perrier out of airplanes," said Andy Stahl, the director of Forest Service Employees for Environmental Ethics. In the first weeks of the Dixie Fire, the air show swallowed over a million dollars each day. One single airdrop of retardant can cost upward of eighty thousand dollars; most of that money goes to the chemical corporations making the retardant, the airline corporations leasing the planes, and the contracted pilots flying them. "The retardant contracts are very profitable for a small group of very influential firefighting companies and for the bureaucracy that uses them," said Stahl.

Other fire profiteers make their money more surreptitiously. Environmental regulations and reviews are loosened or waived during wildfires, which occasionally allows for resource exploitation in the name of firefighting. In one case, a Forest Service official allowed a timber company to cut down a fifty-mile-long, three-hundred-foot-wide expanse of forest, under the guise of protecting a community from the Wolverine Fire. According to an investigation by Forest Service Employees for Environmental Ethics, "The line did nothing to stop the Wolverine Fire." It did, however, "provide timber—a lot of it—without the environmental reviews and public input that otherwise would have been mandatory. The line provided enough timber to fill almost 1,000 logging trucks, some of it massive old growth."

"Their theory is that it's a war," said Stahl, referring to senior Forest Service officials and their contractors. In war, collateral damage is expected. "It's the fire-industrial complex, the nexus between corporate and government agencies combined, with really no interest in ending warfare on wildfires. It's ever-increasing."

If wildfires attract disaster capitalists, the aftermath brings the scavengers. As soon as a fire is contained—before the smoke is out, when the forest is a million black toothpicks protruding from fields of white ash—convoys of bulldozers and logging trucks rumble into the mountains. Most are deployed by billionaire Archie Aldis Emmerson.

Emmerson, the former CEO and current chairman of Sierra Pacific Industries, makes more money from logging after fires than any other person in America. He buys the wood from the Forest Service at a discount, then turns the lumber, 90 percent of which is usable, into boards and wood products for Home Depot, Menards, and Lowe's. "This is a profitable niche," writes Chloe Sorvina, a journalist at *Forbes*. Sierra Pacific Industries has annual operating profits of $375 million. As much as $100 million of that comes from postfire clear-cutting, a practice that removes all the trees from the area that burned.

The term for this practice is "salvage logging." From an environmental perspective, this is often a misnomer. By removing dead trees from burned areas, postfire clear-cutting often disrupts the ecological cycles necessary for regeneration. Heavy machinery churns the soil, damaging any seed banks that survived the fire. The loss of trees—both live and dead—eliminates the habitats of animals that would foster the recovery process by eating and dispersing the seeds. Removing the trees also strips the land of nutrients that would otherwise have been deposited back into the soil to allow the seeds to grow. Much of that nutrient load is comprised of carbon, which, when logged, is released in lumber mills as greenhouse gases.

Postfire logging corporations, however, are concerned not with salvaging the land, but with salvaging their profits. After clear-cutting the land of all trees damaged by wildfire, Sierra Pacific Industries plants new trees, croplike, of whatever species is fetching the highest price. The cycle is self-perpetuating. The resulting homogenous forests are more likely to burn, and with more intensity.

From the first spark to the last tendril of smoke, from the fire camp to the fireline to the helicopter-filled sky, megafires bring in money. In that sense, the industrialists capitalizing on fire suppression are not so different from the hotshots. The economic incentives are in place, and the emotions follow. It is not necessarily a problem that people make a living from fire suppression; fire suppression is needed so long as degraded forests and a carbon-infused atmosphere produce destructive megafires. The problem arises when corporations that profit from disasters accrue enough wealth to gain political influence, foreclosing possibilities of fire prevention and sustainable change. If hotshots are cogs in the machine of the war on nature, industry is the engine that keeps it moving.

As the climate crisis accelerates and the disaster economy grows, so does the influence of corporations that benefit from these same disasters. In 2017, the president of Grayback Forestry gained an audience with Ryan Zinke, the Trump administration's appointee to lead the Department of the Interior. Zinke was an inauspicious character to weigh in on wildfire containment strategies—while in office, he denied the role of climate change in megafires; during his tenure, he was accused of ethics violations for doing business with an executive at Halliburton, a corporation he was charged with regulating. However, the mission of privatizing fire suppression aligned with Zinke's goal of privatizing the public sector, including America's lands, forests, and disaster response.

Chemical companies are winning this race to the bottom. A decade ago, the parent company of Perimeter Solutions was spending an average of $100,000 per year lobbying Congress on issues including "fire suppression and general management of U.S.F.S. lands." They received a solid return on their investment; between 2020 and 2021, the Forest Service spent about $200 million on fire retardant, an industry over which Perimeter Solutions enjoys a monopoly. Not to be outdone, even major aviation corporations have emerged as fire policy lobbyists over the past decade. Since 2009, Mark Rey, who oversaw the Forest Service as a former undersecretary of the U.S. Department of Agriculture, accepted $954,000 from Lockheed Martin to lobby Congress on fire policy.

Logging corporations also wield a disproportionate influence in the realm of forest and postfire management. Emmerson, the salvage logging baron, defends his business practices by saying that the regulations governing timber harvests prevent malfeasance. But his daughter-in-law formerly worked for the Environmental Protection Agency, which is tasked with making the rules he follows, and she now lobbies Congress on behalf of Emmerson's industry. Her lobbying effort, like the others, has paid off. In 2018, the Trump administration approved a $28 million "reforestation" plan that, in classic doublespeak, waived environmental reviews to allow for the clear-cutting of burned areas. Four years later, Emmerson contributed nearly a million dollars to kill a California ballot proposition that would have allocated money to improving forest health to reduce fire hazards. Emmerson, like other billionaires, is investing in the megafires of the future.

Back at fire camp, Scheer wanted to get in on that investment. We emerged from the porta-potties at the same time and washed our hands across from each other. The water made our hands glow against the layer of dirt on our arms. Scheer's hair was greasy and tousled. His cheeks had sunk into his face, giving him a square grin and an unhealthy puckered look. But his brown eyes were earnest as he stamped his boot to produce a trickle of water.

"Yo, Thomas, I have another idea." Scheer sounded drunk. The heat did that to him. "You know how much these porta-potties cost? You can rent them out for *one hundred dollars a day.*"

I was holding my breath against the stench of human excrement, so I didn't reply. Scheer didn't notice. Gazing down the rows of toilets, he was in the thrall of a dream. There were thousands of us fighting fire today. There could be ten thousand tomorrow. And everyone would need somewhere to shit.

"*One hundred dollars a day.* I think I'm gonna go into the porta-potty business."

Part IV

ESCALATION

CHAPTER 11

ON JUNE 5, 1991, A GROUP OF MEN GATHERED IN A CONFERENCE room at the Capital Hilton hotel in downtown Washington, D.C., a few blocks from the White House. The spring day outside was a magnificent 73 degrees, but within the hotel, the men wore blazers against the bite of air-conditioning. They kept the air crisp because they needed to remain focused. They had gathered at the behest of the Cato Institute, a think tank founded by Kansas-based petrochemical billionaire Charles Koch, to form a unified front against the threat of climate change. But the threat they sought to address was not the existential one that faced humanity. It was the threat climate regulation posed to the fossil fuel industry.

The conference was called Global Environmental Crises: Science or Politics?, and it is the first documented event directed explicitly toward climate change denial. The attendees' task would prove a difficult one. For decades, the fossil fuel industry's own scientists had been uncovering alarming results. In 1958, Charles Jones, an executive of Humble Oil and Refining Co., the forerunner of Exxon, first informed the American Petroleum Institute about the potential environmental impacts of increased carbon emissions resulting from fossil fuel consumption. The next year, Edward Teller—a famous scientist who helped invent the hydrogen bomb—explained to executives, "Whenever you burn conventional fuel, you create carbon dioxide, and its presence in the atmosphere causes a greenhouse effect." By 1965, the scientists' warnings grew more pointed. "There is still time to save the world's peoples from the catastrophic consequences of [carbon] pollution," said Frank Ikard, the president of the American Petroleum Institute, when addressing his industry audience, "but time is running out." In 1980, Stanford physicist John Laurman warned the American Petroleum Institute that continued use of fossil fuels would produce "globally catastrophic effects."

Even a memo circulated among Exxon managers observed that their company's long-term business plans could "produce effects which will indeed be catastrophic (at least for a substantial fraction of the earth's population)." The following year, an Exxon report predicted with precision how much the planet would warm and the damage this would inflict. Independently, in Europe, Royal Dutch Shell's scientists produced similar findings. Fossil fuels would cause disruptions that may be "the greatest in recorded history," including abandonment of entire countries and forced migration around the world. "Civilization," Shell's scientists concluded, "could prove a fragile thing." Like the other reports, this one was stamped CONFIDENTIAL.

By 1991, when Charles Koch gathered his men in that air-conditioned conference room in Washington, D.C., the fossil fuel industry had been aware that it was causing climate change for decades. However, for them, the real problem had just begun: scientists at public institutions were urging Congress to act. "We can look down the road a little way and see an industry under siege," said Lew Ward, an Oklahoma oil executive and member of the Koch network. "We are not going to let that happen."

⁂

THIRTY YEARS AFTER THAT FIRST CLIMATE CHANGE DENIAL CONFERence, a heat dome descended on the West. In Canada, temperatures were hotter than on the equator. In Oregon, the heat killed hundreds. In California, we were back in Big Sur.

I never wanted to be back in Big Sur. The beauty mocked me. The forested valleys were almost enough to distract from the sheer cliffs and razor brush. Almost. I saw the mountains as walls. The brush was thick, brittle, and filled with flammable wax. My body remembered these mountains better than my mind: the heat, the rash of poison oak, and the hole that opened up inside me when fifteen colleagues narrowly survived a burnover on the ridgeline above. Now we were back in that terrain in the most extreme regional temperatures ever recorded.

With the hotshots, I worked down an exposed slope—123 degrees Fahrenheit, no shade, shimmering light. This was earlier in the season, before I was taken off the chainsaw. Smitty was my puller. The slope was steep, and we struggled to keep our footing as we cut a line fifty feet wide. Edgar was planning a nocturnal burn operation, if conditions cooperated. That wasn't likely. A mile away, three-hundred-foot flames snapped and recoiled like hellish serpents in a mountain bowl. Aoki said it might be the most extreme fire behavior he had ever seen. It was only June.

Republicans often criticize forest management in order to divert discussion away from climate change. However, when explaining the growth of megafires, forest management is more of a factor in some landscapes than others. In Big Sur, as in much of central and southern California, chaparral landscapes have evolved to burn violently every thirty to one hundred years. If they burn more frequently than that, the young plants die before they have produced a seed bank. This erases the ecosystem and causes a transition to invasive grasslands, which are more flammable and hold less carbon. Axel warned me that if I stayed in this job long enough, I'd feel like I was on a wildfire merry-go-round, returning to the same places over and over again. Axel had fought a fire in this exact location just five years before. This land shouldn't be burning again. Not yet. Just a generation ago, it would have been nearly impossible for a wildfire to burn this early in the year. The fire we faced that day didn't emerge from the legacy of fire suppression alone. This was climate change in its raw form.

Most of the research on climate change and wildfires focuses on large-scale trends. These trends are important. The planet has heated continuously since the early 1900s, when industrial nations began burning massive quantities of fossil fuels. California's temperatures have risen roughly 3 degrees Fahrenheit over that same period. Three degrees Fahrenheit may not sound like much. Most people would not notice if the temperature in a room warmed that amount. But these incremental changes are devastating when applied across landscapes. Hot air is like a dry sponge, sucking water from everything it touches. The warmer the air, the drier the vegetation, which means that even small increases in average heat cause major changes in plant flammability. Coupled with earlier springs, diminishing snowpacks, and longer droughts, dry vegetation causes California landscapes to be more flammable for more of the year than they have ever been. In just fifty years, climate change has extended fire seasons by nearly three months and expanded the state-owned land area facing severe fire risk by 15 percent. Scientists have concluded that climate change is now the most significant driver of extreme fire behavior in the western United States. That wasn't supposed to happen for decades.

Yet climate change does not only unfold as a broad trend; it is punctuated by violent outbursts of extreme weather events. Firefighters know that weather is the main factor that drives real-time fire behavior. While we were cutting, Jack stepped aside once every hour to calculate the weather in our exact location. The mechanism was archaic, a thermometer attached to a string that he dipped in distilled water and spun for sixty seconds to calculate the temperature and humidity. He broadcast his results on the radio, which allowed us to estimate the amount of moisture in the vegetation around us. While California's vegetation is drying out on broad scales of time and space, heat waves accelerate everything, sucking out whatever moisture is left. These heat waves are already several times more likely than they were a half century ago, when atmospheric carbon levels were within their historical average. In California, heat waves act like a powder keg, making fire behavior exponentially more extreme.

That day, I felt climate change as a physical reality. The same dry heat that had turned plants into tinder now pulled the water from my body. What little moisture remained poured from my skin to keep me from overheating, leaving nothing to keep my muscles hydrated. My sunglasses were caked with sawdust. If I cut too much, I would waste valuable energy. If I cut too little, our prescribed burn could cross our line and kill us. And the calculations were infuriatingly subjective—I was supposed to "eyeball" it through sweat-stung eyes as we chugged down the mountain.

I felt my cognition slipping. I stopped cutting for a moment, got my puller's attention, and made a chopping motion with my glove toward a prominent manzanita bush ahead to ensure we were still on the right path.

Smitty didn't want to collaborate. His arms went akimbo and his face contorted. "Stop second-guessing yourself and just fucking cut!" he screamed.

I began to reply, but his expression gave me pause. His eyes were unfocused, bloodshot, already addled from the heat. He looked drunk.

He was angry, and I couldn't blame him. Every bush I cut, he had to drag fifty feet across the slope and throw atop the rest. The closer he came to passing out, the angrier he got.

I was angry too. Angry at my puller for being a jerk. Angry at the fire, at the heat, and at the pain. Angry that it was June and temperatures were double what they should be. Angry at the men who sowed the seeds of climate change denial. My chainsaw was screaming. I let my anger slip into it as branches rained around me.

※

BY THE TIME THE MEN STEPPED OUT OF THE AIR-CONDITIONED HOTEL in Washington, D.C., they had a plan, but they faced an uphill battle. In the early 1990s, when the threat of climate change was well established, the need for policy action was accepted across party lines. "Those who think we are powerless to do anything about the greenhouse effect are forgetting about the White House effect," said George H. W. Bush, the Republican president at the time. Even John McCain was concerned; he introduced a bill to impose an economy-wide limit on carbon emissions.

Bipartisan political action seemed imminent. In 1972, after scientists discovered that a prominent pesticide was decimating ecosystems, the Environmental Protection Agency banned it. In 1988, after scientists discovered a growing hole in the ozone layer, Ronald Reagan banned the chemicals causing it. In 1990, after scientists traced acid rain to sulfur emissions from coal-fired power plants, George H. W. Bush strengthened the Clean Air Act, which was originally signed by Richard Nixon. Climate change seemed poised to follow a similar path, nipped by legislation early on. Clean-energy technologies were in their infancy, but with public subsidies to trigger market innovation, they could replace fossil fuels within decades. The path was economically, technologically, and politically straightforward.

The fossil fuel industry was determined to change that. Charles and David Koch were already veterans in this task. The Koch brothers' father had provided Adolf Hitler with an oil refinery in the lead-up to World War II, and he had so admired Germany at the time that he hired a fervent Nazi to nanny the brothers in their youth. When the Koch brothers inherited their father's empire, they turned it into a political weapon. They helped fight the civil rights movement in the 1970s. They helped Big Tobacco fight regulation through the 1980s. They were motivated, according to biographer Christopher Leonard, less by overt greed than by an earnest, if infantile, belief that regulations imposed on multinational corporations by democratic societies are a slippery slope to communism. While the Koch brothers called themselves libertarians, Harvard scholar Naomi Oreskes argues that they warped this term beyond recognition, creating a pseudo-religious creed of fundamentalist capitalism that is as unscientific as it is unwavering in its extremism. And a zealot is the most dangerous kind of villain.

By 1991, when the Koch brothers began their campaign against climate science, their strategies were already well established: attack the research, attack the scientists, and, when that fails, attack the process of science itself. This was the easy part, appropriating the tactics used by the asbestos industry, the lead industry, and the tobacco industry. Aided by the birth of Fox News in 1996 and assisted with millions of dollars from Exxon, the Kochs painted their propaganda machine with a scientific facade. Their proxy groups—the Cato Institute, George C. Marshall Institute, Heritage Foundation, and many others—manufactured controversy where none existed. They exaggerated scientific uncertainties, criticized sound climate models, and championed counternarratives such as global cooling. Doubt was their product.

The fact that this manufactured doubt disproportionately influenced Republican voters was no coincidence. Amid the swirl of misinformation, the Koch team deployed its influence—specifically in the Club for Growth—to fund election challenges against any Republican member of Congress who demonstrated a modicum of scientific literacy or social obligation related to climate change. Since 2000, approximately 85 percent of the fossil fuel industry's campaign donations have flowed to the Republican Party, with the Koch network alone spending hundreds of millions per election cycle. Even John McCain flipped his position on climate change after he was nearly unseated by a Koch-funded denialist. By 2014, the fossil fuel industry had transformed climate change from a bipartisan issue to one that only 8 of 278 Republicans in Congress were willing to acknowledge was real.

Climate change denial proved lucrative. In just the three decades since that first climate change denial conference in the air-conditioned hotel, during the years that fossil fuels should have faded into obsolescence, more fossil fuels have been pumped into the atmosphere than in the entire history of humanity before. In those same years, the industry has enjoyed record profits, averaging approximately $3.2 billion in profits every single day. As our bodies began to collapse in Big Sur, atmospheric carbon levels were higher than they had been in three million years.

And the temperature just kept rising.

❦

BY AFTERNOON, I COULD FEEL MY BRAIN BEGINNING TO OVERHEAT. MY heart was racing like it was going to explode. My vision began to close, red shadows spreading from the edges until I was looking through a pinprick of blinding light. It took all my focus to hold on to that light. I knew that if I allowed it to close, I would be finished. I would pass out. If I survived, I would wake up in a private medical helicopter. I could not afford the flight. The light was what was keeping me conscious, so I focused on it with all my will, relegating the chainsaw to the instincts of my body.

My work became sloppy. My chain was dull, making each cut more difficult. I killed the engine, dropped to my knees, and opened the chainsaw with a wrench to swap out the chain. Edgar watched from above, perched on a sandstone protrusion like an overlord. The only thing I could smell was gasoline. I blistered my thumb against the chainsaw through a hole I had worn in my leather glove. Sweat dribbled from my chin and sizzled against the metal. One drop, then another. I wasn't sweating enough. Chills crept along my skin. Some corner of my mind that hadn't yet fried whispered that these were signs of impending heatstroke.

I was too tired to feel afraid. Too hot. I moved mechanically, still on my knees, pulling a canteen of water from my pack and dumping it on my head. I almost cried out. Maybe I whimpered. The water pouring down my back was the temperature of the air, nearly hot enough to scald me.

I stood. My vision was a throbbing mass of black and red, closing around the last point of light. I focused on that point of light. *Stay conscious.* The chainsaw was screaming again. I must have turned it on. Life was crashing around me, wood and wax and leaves and the last salt from my own body. Eventually I couldn't hear through the pounding of my pulse, couldn't feel the sticks and twigs cutting into my skin, could no longer feel my own body. I cut a low stump, chainsaw teeth zipping inches from my boot. *That's all it would take*, a little voice whispered inside my head. *Just one nick.* At this point, I wouldn't even feel it. My body was far past the realm of sensation. *One toe. Just take one toe. Then this will all be over.*

Shit. I had blacked out. When the pinprick of light opened again, I was standing, my legs wobbling like jelly, cutting a rock at full tilt on the edge of a small cliff. Stone on metal, sparks flying. I didn't worry about the sparks. The chain was ruined again. This jolted me back to my body. Edgar was watching. If he saw me swap another chain, he would be angry. I couldn't handle his shouting. Not now. I would break.

Before I could regroup, the rocks moved under my feet. My balance had fled long ago. The ground beneath me disappeared. I heard Smitty shout in the distance, but I was already in the air, body limp, plummeting with my chainsaw down the cliff.

AS MY BODY COLLAPSED ON THAT RIDGE, THE HEAT WAS CAUSING power cables to melt and roads to buckle all along the West Coast. In cities, paramedics ran out of cooling stations, so they resorted to filling body bags meant for cadavers with ice, then zipping unconscious victims of heatstroke inside. When people collapsed on sidewalks, they suffered third-degree burns. In hospital records, the medical thermometers designed to read the body temperatures of victims all came out at 107 degrees, which was mysterious until doctors realized that the instruments were not designed to go any higher. Leila Carvalho, a meteorologist, told me that atmospheric scientists faced a similar problem—their current instruments weren't equipped to monitor the temperatures we were working in.

For many years, scientists mistakenly believed that climate denial was based on a legitimate disagreement over scientific data. Scientists labored to fix this, honing their methods and models until the link between fossil fuels and climate change was about as airtight as the theory of gravity. During those same years, renewable energy technologies advanced through the headwinds of opposition, becoming, in most cases around the world, the cheapest and most reliable fuel sources. By 2021, when scientists at Princeton released a report charting five pathways to net-zero carbon emissions using existing technologies, the technological and financial barriers to a transition away from fossil fuels had been greatly reduced, if not eliminated.

For a moment, the tide seemed to be turning. At the end of his first term as president, Donald Trump, who before had called climate change a "Chinese hoax," could no longer deny its reality. Most Republican voters followed suit; by 2021, 61 percent finally accepted the legitimacy of the phenomenon. Even Ben Shapiro, the former editor at large of the alt-right Breitbart News and the most listened-to conservative podcaster in the United States, championed his own conversion from denier to believer. But any pretense that the disagreement had ever been about facts crumpled when these converts settled on the same prescription as before: keep burning fossil fuels, no matter the cost. "The question isn't whether climate change is happening," Ben Shapiro tweeted in 2020 as smoke from California's fires reached Europe. "It's whether you have any solutions that aren't crazy."

Ben Shapiro provided a megaphone for the fossil fuel industry's pivot from denialism to delayism. Instead of arguing against the science of climate change itself, this new form of denial works to downplay the threat while hobbling proven solutions. Delayist tactics are just as unscientific as outright denial, and perhaps even more dangerous, because they allow Republicans to adopt an air of false pragmatism while propping up the industry that is scorching the earth.

This new line of attack is funded by the same people who funded the old one. In 2018, five of the largest fossil fuel corporations spent $200 million lobbying to control, delay, or block climate policies. The same industry front groups that once flooded the media with climate change denial simply shifted the target of their misinformation, inundating the public with claims that solar farms shed toxic chemicals into the water supply, that wind turbines caused cancer, and that the energy transition was a Trojan horse for communists intent on stealing individual liberties.

The current strategy, to use the words of Steve Bannon—who worked with Ben Shapiro at Breitbart and is a former Republican White House political tactician—is to "flood the zone with shit." By inundating the public with a constant stream of misinformation, fossil fuel operatives make scientific facts appear unknowable, which paralyzes action and deflects legitimate debate. Fossil fuel pundits now justify their opposition to offshore wind with a plea to save the whales. The government of Tennessee officially labeled methane gas, which produces planet-warming particles twenty-eight times more potent than CO_2, as renewable energy. And Exxon styles itself as an innovator of climate solutions by championing carbon capture, which the Intergovernmental Panel on Climate Change (the world's most authoritative body of climate scientists) clearly describes as "unproven," the reliance on which would present "a major risk in the ability to limit warming."

Misinformation often takes on a life of its own. In 2018, during weeks of wildfires that cloaked the whole West Coast in perpetual amber dusk, Majorie Taylor Greene, a Republican congresswoman, blamed Jewish space lasers for the unfolding destruction. Others believed that the government of California started the wildfires to cull a population already vulnerable to a respiratory pandemic. In 2020, right-wing militias in Oregon began showing up at wildfires with assault rifles, claiming that the conflagrations were lit by leftist activists. All of this is enough to make conventional, dinner-table climate change denial seem moderate, until you realize that these conspiracy theories are merely a different means toward the same end: protecting and prolonging an industry that is killing us.

Some of us, that is. Climate change is killing some of us. But for a thin sliver of the American population—those with enough wealth to insulate themselves from the heat and enough willful ignorance to overlook the ethical fallout—climate change still appears to be a distant threat. The effects of climate change land hardest on the world's poor. Our current trajectory "will not avoid the death of millions," writes climate scholar Laura Pulido, because, for people profiting from the destruction, those millions "simply do not matter."

Until I became a hotshot, I had imagined the lines separating the masses from the elites to be the thin film of a bubble's edge. When climate change struck closer to home, I thought perhaps the bubble would burst, forcing a reckoning with fossil fuel damage. Maybe when firefighters burned in a local forest, or mansions were washed away in mudslides, or climate refugees appeared not just at our borders but within our own communities—maybe then the bubble of "us" versus "them" would burst, forcing an ideological realignment. Instead, I saw that bubble of valued humanity shrink, not expand, when climate disasters hit home.

When this reality set in, Ben Shapiro's words made sense to me. "It's hard to call something a 'crisis,'" argues the subtext to a talk he delivered within an air-conditioned auditorium, "when the climate debate has lasted for multiple decades and *we've* yet to see the effects" (emphasis mine).

I WAS IN THE AIR, FALLING. I FELT MY BODY TURN UPSIDE DOWN. MY helmet hit stone. The world slowly rotated.

I cartwheeled. I saw smoke. I saw sky. Then my head hit stone again.

I often feel that time is elastic. Measuring time by its length, I forget to attend to its depth.

Moments shift.

Sometimes they slip, hours lost in patterns of breath and pulse and movement.

Sometimes they slow, their hidden dimensions stretching, sinking, and falling away. I had felt this distortion before, in love, when Kenzie first fell asleep in my arms. Those ten minutes before she woke could have been a lifetime. And I seemed to fall into that space now, in the air, flying down that cliff, feeling each beat of my heart.

My face was approaching the chainsaw, wedged against a bush below. I could not alter my trajectory. I knew what the teeth would do. Twenty-four razors spinning twelve thousand times per minute, carved to bite and spit matter with each rotation. I had a tourniquet in my pack, but it was meant to clinch blood from a severed limb, not my head. Had I dulled the teeth enough to save myself? Was my skull harder than ironwood? Probably not.

At a certain point of exhaustion, the layers of consciousness fade, alongside the feeling of significance we carry in daily life. With whatever awareness remains, you understand that your relationships, your sense of meaning, your hopes and fears and dreams are woven into a quilt that comforts your self-aware mass of gristle and bone. If that quilt is pulled back, you feel like nothing but a jumble of flesh and sinew within a gnashing world—the same world that fills our lungs, becomes our blood, builds our bones, and takes them back into the earth. There is a connection in that loss of self. There is a bliss when the pain becomes so intense that the planet takes you and the shock of insignificance fades.

So I felt nothing in my own body as the chainsaw grew in my vision. I heard the rattle of rocks, the tumble of brush, a roar of blood in my ears. My helmet hit the chainsaw's teeth. I heard a crunch. I still felt nothing. My neck slid down the chainsaw blade. Still nothing. I fell another ten feet in loose stone, headfirst, until I came to rest against a bush.

I lay under that bush, not moving. The shade felt nice on my face.

Maybe I would sleep here. I could finally rest, just for a moment. I had fallen from a cliff, so I had an excuse.

I don't know how much time passed. It couldn't have been long, because I was still alone when I wiggled my fingers and toes to ensure I wasn't paralyzed. They moved. Then I flexed my muscles to see if any bones were broken. No pain. I flexed again, just to be sure. I felt blood on my chin. Blood should be hot, but this blood was cooler than the heat radiating from the rocks around me. I would have a black eye, but my body was intact. I must have instinctively hit the saw's chain brake with my forearm when I fell, locking the teeth. Without that instinct, I might be dead. I congratulated myself. Maybe I wasn't the kook Scheer said I was.

I could see Edgar observing me from far above. My mind went to my ruined chain. Edgar had seen me fall. My few functioning brain cells hatched a brilliant plan. I would lie to them. I would *lie*. I would say the chain hit a rock when I fell. No one would blame me for that. I could swap the chain again without getting yelled at. *Of course* the chain was damaged when I fell from the cliff. I had been cutting like a real hotshot before I fell. Lying there, hatching my plan, I was thankful that I fell. So thankful. Thankful for the rest. For the alibi.

I stood, wiping blood from my chin. I climbed back up the cliff and swapped the chain. No one yelled at me. Perhaps no one noticed. Everyone was too hot to notice. I got back to work.

WHEN I REJOINED THE HOTSHOTS AFTER MY FALL, EVERYONE WAS IN bad shape. Drogo later told me he considered throwing himself in front of Scheer's saw to escape the heat. Scheer, for his part, tripped onto his chainsaw, almost gutting himself. Another sawyer briefly lost his mind, muttering aloud as he crouched on a boulder to deconstruct his entire saw. When the metal burned holes in his leather gloves and blistered the palms of his hands, he cackled, high-pitched, keening.

There were no fixed climate politics on my hotshot crew. Most, like Aoki, Axel, Scheer, and Barba, were firmly on the side of science. Others didn't care. Márlon told me he couldn't give less shits; he had no political power, he just needed to pay his bills. But the day of that heat wave, as the sun began to set and we finished our line, I asked Jack about his thoughts. Through a beard crusted with salt from his sweat, he told me that Democrats and billionaires were controlling the weather to justify a socialist takeover.

Jack's stance surprised me. It felt self-destructive for him to express solidarity with a fossil fuel industry that was threatening his own life. But iterations of his view were held by others on the crew who harbored a deep suspicion of the same federal government they worked for. I came to view theirs as a different category of denial. While market fundamentalism made sense to me as a craven dogma held by petrochemical billionaires and other beneficiaries, the denial of the hotshots felt like something else.

Throughout the season I noticed that, while some hotshots would deny climate change during banter, they would offer much more nuanced viewpoints during earnest conversations. From the perspectives of many hotshots, belief in climate change and knowledge of its causes wouldn't change anything, politically or otherwise. Rather, it would simply signal an identification with a federal government that employed them, but which they had learned through experience to distrust. I found I couldn't blame them. For many who feel disenfranchised, the voices clamoring for climate action appear to be the same amorphous elites who refuse to provide them with living wages or reliable health care. From the outside, the logic seems twisted, but from the inside, it feels immaculate.

What surprised me wasn't the hotshots' suspicion of institutions, but the fossil fuel industry's ability to turn these sentiments to their own advantage. By trying to speak truth to power, some hotshots were unwittingly propping up the corrupt elites they felt they were opposing. Their political nihilism was just another commodity for the fossil fuel industry to capture and exploit.

By the time we reached the stream, another hotshot crew was in the water. "C'mon!" one shouted. "Get in! It's so good!" We all looked to Red. He pursed his lips and shook his head. He was afraid Edgar was watching. We needed to maintain appearances, even now. We needed to be hard.

Márlon laughed at Red, took off his shirt, and limped to the other side of the stream, where he lay in the mud so that the whole top half of his body was submerged. Red sputtered a protest as everyone followed suit. I stretched my legs along the bank to avert muscle cramps, knelt, then plunged my head into the water. The sounds of dusk—laughter, voices, crickets, birds—were cut off as my head entered the stream, replaced by the white noise of the current. I held my breath, opened my eyes to see particles of dirt flowing off my skin in the green light. I wanted to stay here, in the water, out of the heat and stress and confusion of it all. When I pulled myself to shore again, I saw that the sun had dipped behind the mountains. I realized that the heat of the day had leached the color from the land. For that brief hour of dusk, the forest drifted into a deeper green, the sky a truer blue.

I noticed that Axel wasn't in the water. He was puffing a hand-rolled cigarette, watching the tower of smoke with a concerned expression. I went to stand by him. This wildfire wasn't going to fade with the night. There was no telling when it would go out.

Axel threw an arm around my shoulders, drunk with exhaustion. "Behold," he said. "Behold the dragon."

CHAPTER 12

On September 10, after suppressing our fourteenth wildfire of the season, the Los Padres Hotshots broke for three days off. I caught a ride home with Barba, with whom I'd become tight. After my demotion from sawyer to puller, everyone seemed to feel that I was in my proper place. The hotshots invited me to smoke cigarettes after dinner, included me in stories, and laughed when I almost blew up our propane stove in Yosemite National Park. The increasing severity of the conditions we faced made this intimacy invaluable.

Kenzie was using our shared Prius. Barba thought that was funny, so he offered to drive me home. We cruised out of Paradise Canyon, wearing fresh shirts for our return. My other two shirts, which I'd worn for a week each, were scrunched into my duffel bag, exuding the metallic smell of atrophy. By this point, our bodies had long since burned through our fat and were now consuming muscle.

As we crested the mountain pass, Santa Barbara appeared three thousand feet below, nestled between the ocean and the foothills. Five months into our season, the sight still conjured a longing: Kenzie, my dog, and my bed were somewhere down there in the blue Pacific mist. Barba tapped his rearview mirror.

"See that?" he asked. I looked back. There were thunderclouds rising from the jagged horizon behind us. Their lightning would start new fires. "I'll bet you a thirty rack that's where we'll be headed in a few days."

"That storm's a long way off." I ignored the bet. As a new guy, I provided the crew with beer anyway. "Think it's over the sequoias?"

"Let's hope not."

Hotshots hate fighting fire in the sequoias. The terrain is steep, rocky, and hot. Until fires burn into higher elevations, hotshots spend their days tangled in whitethorn, a brutal plant that leaves the tips of its spikes in their skin until the lesions swell and exude pus. Even worse, there is no phone service in those mountains, so time drags.

My days at home were their typical blur. I slept late, walked my dog on the beach, had dinner with Kenzie, and showed my face in the anthropology department to remind them I was alive. Hotshots called this the "real world," but it felt surreal. On the third day, the anxiety began to mount. I found myself answering Kenzie's questions with robotic crew language: "affirm," "negative," and "copy." When I tried weeding the garden, I entered tunnel vision and upturned the entire plot like I was digging line. I retreated to menial tasks: laundry, cleaning the dirt and soot from my boots, washing the sweat stains from my hat. I wanted to get back to my crew.

In the evening of my third day at home, I got the text from Red.

"Got an order to Porterville. Wheels rolling at zero-six-hundred." Porterville, in hotshot lingo, means there is a fire in the sequoias.

The next day, my crew drove through coastal mountains in a foggy dawn, crossed the Central Valley between oil derricks and industrial farms, and climbed into the Sierras. We found the fire puffing quietly in a valley of oak trees. While Aoki scouted the attack, I stood by Márlon. We watched the smoke from our perch on a grassy knoll.

"Think we'll be here awhile?" I asked.

"Nah," Márlon said. "We should have this thing wrapped in about four days."

I hoped so. Above the smoking valley, mountain slopes cut toward the sky, rising into peaks and ridges under a blanket of pine trees. The vegetation was so dry it crackled. In low elevations, the forest was brown and dead from drought. On the highest ridges, clusters of sequoias rose above the rest, silhouetted against the blue sky.

The sight of the sequoias reminded me of a safer time and place. During the first pandemic lockdown eighteen months prior, on the other side of this same ridge, Kenzie and I had been alone in a sequoia grove. We climbed into wooden crevasses the size of cabins while our puppy galloped around the trees, barking and wagging her tail. For me, the sequoia landscape had become personal. It was impossible not to infuse these trees, among the oldest organisms on earth, with a meaning that transcended words. I understood why images of sequoias being logged for lumber had spurred the conservation movement a century ago, and why Indigenous people have recognized their significance for many thousands of years.

❧

THE GIANT SEQUOIA, OR *SEQUOIADENDRON GIGANTEUM*, RESEMBLES something from an older world. Up close, each sequoia is beautiful. Its bark is the tone of warm sunset, grooved, soft and spongy to the touch. Its tall branches filter sunlight like cirrus clouds. But I have always thought sequoias are better viewed from a distance that brings the whole grove into view. From this perspective the grove appears as a single organism, dozens of pillars lifting a shaggy canopy into the sky.

The first sequoias appeared around two hundred million years ago. When the continents were still conjoined and shook with the footsteps of dinosaurs, sequoia forests covered most of the northern hemisphere. They survived the asteroid strike that drove most species extinct, then drifted with the land as continents split. A mere hundred generations of these trees would track the entire history of the human species. Many of today's living sequoias were young during the time of Christ.

Sequoias have evolved to thrive with fire—to rely on it, even. Their bark is thick, with a high water content, which protects the interior from heat. Their cones are serotinous. This means that the high temperatures of flames cause them to open, releasing their seeds into ash-enriched soil that has been cleared of competitors. As sequoias mature, they also shed their lower branches, protecting their foliage from all but the most extreme fires.

And yet all these adaptations would be insufficient for a sequoia standing alone. They rise as tall as twenty-five-story buildings, but their roots only penetrate about twelve feet into the ground. The sequoias can reach such heights because their roots intertwine and sometimes fuse with one another (a unique process called root grafting) to create a continual web beneath the surface. Often, the web spreads for miles, allowing the trees to share water, resources, and stability against uprooting. Without a grove to offer support, a single firestorm would be enough to topple a tree. But the roots don't just support the trees. They also connect them. A sequoia can survive when up to 90 percent of its canopy burns, in part because the rest of the grove can help support the injured tree. The interconnected trees share nutrients, which are transferred through the roots, up the vascular system, and into the limbs, providing the energy needed to heal and regrow.

I was far below the groves, cutting through a field of brush with my chainsaw, when Aoki radioed that we needed to pull out. The atmosphere had opened, the smoke had boiled into the sky, and flickering lights had appeared on the ridge across from us like torches held by an opposing army. The fire was blowing up again.

I walked behind Márlon as we returned to the buggies to regroup. "Wrapped in four days, huh?" I asked.

Márlon grinned. "Looking more like four weeks now."

This time, Márlon was right. The sequoias above us felt like a world apart, a world that was solid, constant, ancient—immortal, perhaps. From a distance, viewing them as a grove, it was easy to forget their vulnerability. It was easy to forget that, in their two hundred million years of life, they had never inhabited a planet like the one we have created.

❧

WE REENGAGED WITH THE FIRE FROM A DISTANCE. HILLS ROLLED toward the horizon, fading in shades of blue. A bulldozer pushed a line of dirt up the ridge. We followed it, working off the dozer line to clear a fifty-foot stretch of brush from beneath an oak canopy. If everything went as planned, we would set a controlled fire to whatever leaf litter was left over, slowly carrying our fire until it burned against the wild one. All the fuel between the dozer line and the wildfire would be consumed, so the wildfire would be contained. The pattern was familiar. The tactics were implied.

What wasn't implied, for me, was that I would remain in the good graces of the crew. I was back on the chainsaw. Scheer was training as the crew leader, so as his puller, I was granted the honor of resuming my tenure on the machine of plant butchery. This frightened me. The good news (for me, at least) was that the crew was still reeling from our COVID outbreak, so there wasn't much opportunity for criticism. Red was back. So was Jack. Jack had been so close to death in his quarantine motel room that he kept 911 on speed dial, but the experience had only seemed to reinforce his libertarian resolve ("Did I die?" he growled at me through his beard when I asked if the experience had changed his views). Several others, however, remained absent. This included Edgar, who, to our collective relief, had thrown out his back moving a sewing machine in the station while recovering from the virus. Drogo bestowed upon the sewing machine the first ever hotshot employee-of-the-month award.

The tattered state of our crew was good news for me because we had two temporary members to fill in for the missing bodies, and the newcomers drew most of the omnipresent animosity. The hotshots called both newcomers Phil. This seemed strange, because I knew one of them as Tim. I had worked with Tim my first year on the beginner crew. Tim was my first puller, and my friend. He was short but muscular, built like a bullet, a unique breed of hippie MAGA Christian I have only ever encountered in California. He had a quick sense of humor, an earnest thoughtfulness, and a gung-ho attitude. After the beginner crew, Tim had joined an engine when I joined the hotshots. Whenever I saw him, I encouraged him to hike hard so he could get an assignment like this one. For Tim, this was a dream assignment because he would have priority getting hired next year as a hotshot. Unfortunately, he attracted everyone's ire when he began puking uncontrollably off the dozer line that first day.

We were all surprised by Tim's vomitous state. This was easy work for us. The slope was gentle, the brush was light, the temperature around 100 degrees, and the fire was miles away. Moreover, Tim looked healthy. His abdomen still had a six-pack. In contrast, by this point in the season, we had become mutants. Our chests were concave, our faces sallow, our skin translucent from constantly being covered up, and our muscles lopsided from disproportionate use. Sawyers' left arms and shoulders were thin, while our right sides were rippling and bulbous. But as repellent as our physique would appear on a firefighter calendar, our bodies were perfectly adapted for the work. Tim's beach body evidently was not.

Scheer came trotting down the line, seeming to enjoy his new role as crew-boss trainee. He carried a radio and a Pulaski, a hybrid ax and hoe. He called for a lunch break, which was a relief, because if Red was still in charge, he would tell us to eat while we worked. I joined Scheer under an oak tree. We watched Tim shiver and dry heave as Axel dribbled water on his head. Scheer's face conveyed disgust, like he had caught Tim engaging in some act of terrible perversion.

"Friggin' Phil," Scheer muttered. I asked why everyone called him Phil when his name was Tim. Scheer glared at me sideways as if I had joined in on the perversion. "They don't have *names*," he said. "They're filling in. It's 'Fill 1' and 'Fill 2' unless they join the crew. Then it's 'New Guy' until they show they can hike."

"Ah. Gotcha. 'Fill.' Not 'Phil.' My bad."

Mystery solved. I felt bad for Fill since I had encouraged him to try to get on the crew. On that first day, he was already as good as dead, socially. But I didn't have time for sentimentality. I had my own reputation to worry about, and it was time to get back to work.

I FOUND RED STANDING IN A PILE OF CHURNED EARTH. HE HAD SPENT the day guiding a bulldozer up the ridge, tying neon tape in thickets of manzanita, chamise, and sage so the bulldozer knew where to uproot life, crush it, and push it aside. A line of dirt as wide as a four-lane highway stretched in Red's wake. The rest of us followed with chainsaws, removing whatever plants clung to its edges.

Red looked disturbed, which confused me, because this was all part of the plan. With the wildfire now raging, Aoki had repositioned us to a location he estimated would be seven days ahead of its spread. All we could see was a distant plume. We began building a seven-mile line that would stretch from the low-elevation oak savannahs to the highest sequoia-shrouded peak. Other crews were moving to meet us from the opposite direction. When we met, our fireline would be over a dozen miles long, creating a lasso around the wildfire's head. Then we would fight fire with fire, burning everything between.

There was nothing unusual about this plan, but when I sat by Red during a five-minute water break, I saw that his face was flushed, mouth downturned, eyes darting. He always spat more and worked his jaw when he was uncomfortable. I asked him what was wrong. Red jerked, like I'd startled him.

"Huh? Oh, what's up, Thomas?" The air shimmered off the metal of the bulldozer. "Wrong? Nothin's wrong. Just got this feeling, you know?"

I thought I knew. Premonitions are common on hotshot crews. They are to be taken seriously. I knew more than one firefighter who had turned down an assignment because of a gut feeling. A friend quit the crew because he kept having nightmares of dying in grisly ways. These premonitions weren't superstition; our bodies sensed things that we might not be able to quickly rationalize, so we knew we should listen when our bodies told us something was wrong. Especially in the last months of the fire season, we were all focused on getting out alive. But Red was among the most stalwart men of our crew. I had never seen him express misgivings.

"Bad juju," Red muttered to himself. "It's bad juju doing this up here." He was squinting down the bulldozer's destructive path. "I don't like it."

"No?"

"No, man. Not here. Bad juju to be doing this here."

"Why?"

"Not our land."

Red told me he was disturbed because the line we were plowing cut directly through the Tule River Reservation. The reservation, eighty square miles of land abutting Sequoia National Forest, is home to the Yokuts, Yowlumne, Wukchumnis, Western Mono, and Tübatulabal tribes. In the Southern Sierra Miwok language, the name for the tree reflects the sound made by the spotted owl, the guardian of the forest. Those sequoias had seen the history that dwelled here. A moment ago, in sequoia time, this had been a river valley that fed the largest freshwater lake in the American West, and the tribes had burned all of it, the right way. A moment ago, in sequoia time, Chumash refugees had trickled into this valley from the Santa Barbara coast, fleeing the Spanish missions with terrible tales. The tribes here had welcomed the Chumash, and they had listened to their tales, so they had known to defend their territory when, a sequoia-moment later, miners and ranchers swarmed. The tribes defeated the Americans in battle. They held the land around the lake, creating a final buffer between the sequoias and the ripples of industry. Then the Americans came in waves, pulling howitzers, spewing bullets, with authorization to take life indiscriminately. The tribes fled into these mountains. They found refuge in the sequoias, the same ones above us, the guardians of the forest.

I didn't know what to say, sitting by Red in the shimmering air of the bulldozer. I often felt this unease. All of California's land is Indigenous land, but the violence felt concentrated here. It felt poignant, as if all the destructive powers of history had accumulated in this small stretch of woods. The lake was gone now, drained for industrial agriculture. Less than 10 percent of the tribal people in this area had survived the genocide. They had rebuilt here, on this reservation. Now, for all we knew, we were bulldozing their sacred sites.

Red stood up again, shaking himself. We didn't have a choice. We had seven days to stop the fire. I went back to my chainsaw, fired it up, and got back to killing.

WHEN JOHN MUIR FIRST VISITED THESE WOODS IN 1874, JUST A FEW years after they were stolen from the tribes, he wrote that they were the finest he had ever seen. I was too tired to notice. For me, the days whirled together in dirt and sweat and chainsaw grease. I measured time by the changing vegetation that fell under my saw, from brush to oak to pine. By the end of the day, maybe it was the third day, we drove through the subdued glow of a forest at dusk. We were approaching a new campsite. The stars would be bright up here, the air cool.

Then a darkness enveloped my window. It came suddenly, like an eclipse, jolting me from my reverie. The darkness was not space, but solid matter—grooves and wood and moss. We drove alongside a fallen sequoia as if through a tunnel, three hundred feet along the road. Its girth was higher than our roof. To enter our campsite, we passed through a cross section that had been cut from its trunk so that, for a moment, we were enveloped by tree rings, hundreds of them, together mapping a single life.

According to most accounts, the first white man to encounter a grove of giant sequoias was named Augustus T. Dowd. In the winter of 1852 he was hunting for meat to provision a mining crew when he accidentally stumbled into one of the largest groves in the area. Dowd described the encounter like a dream. The sequoias swallowed the sounds of the forest, red bark glowing against snow. They seemed to prop up the sky itself. Soon settlers were calling the most impressive giant of the grove the Discovery Tree. They reckoned the Discovery Tree was more than one thousand years old. As soon as the snow melted, the settlers cut it down.

They began by skinning the Discovery Tree, stripping the bark to a height of fifty feet. Without bark to protect the phloem layer, which transports nutrients, the sequoia began to starve. Then the men attacked the trunk with pump augers—long metal drills roughly two inches in diameter. They bored dozens of holes, penetrating to the sequoia's core before moving in with saws and wedges, slicing and pounding and cutting for days on end. Finally, on June 27, 1853, after twenty-two days, the Discovery Tree creaked, split, and fell, shattering like glass against the ground. At 280 feet tall, 24 feet in diameter, this was the first sequoia to ever fall by human hands. It would not be the last.

While later visitors would view sequoias with reverence, the overwhelming western mentality at the time of their "discovery" reduced them to objects that were valued only for the profit they could produce. Specifically, sequoias were valued as a potential source of exhibitionary revenue. After felling the Discovery Tree, the settlers moved to its neighbor, Mother of the Forest. Instead of cutting this tree, they girdled her, removing her bark up to 116 feet high and shipping it to New York City, where it was reassembled in the Crystal Palace to demonstrate the supremacy of western culture. Without bark, the Mother of the Forest reportedly starved for seven years before dying. A visiting writer mourned her passing. The tree, he wrote, which once stood as the "proudest of the grove," had been reduced to "a colossal skeleton, holding up its withering arms in silent accusation against the vandalism of its executioners."

In the following decades, as corporations gained control of the mountains, the destruction of the sequoias accelerated. Urban exhibitions of dead trees failed to fetch much profit, so the trees' value shifted to the amount of lumber held within their bodies. One tree holds enough wood to build seventy-five houses. Or, as Southern Pacific, a railroad company, advertised in 1907, "enough lumber . . . to supply a line of poles from Kansas City to Chicago." However, because sequoias shatter when they strike the ground, most of the wood was only good for fence posts and grape stakes.

Nevertheless, the settlers kept cutting. When John Muir later roamed these mountains, he lamented that the sounds had shifted. Where once he had listened to the rush of wind through high branches, the land now reverberated with the racket of lumber mills, "booming and moaning like a bad ghost." By the dawn of the twentieth century, settlers had killed more than one-third of the giant sequoia trees.

As we entered the magnificent fold of the survivors, it was easy to forget that their existence had ever been threatened. In the coming days, we were expected to save the rest. The glow of the wildfire was still far off. This allowed us a rare moment of peace. Some hotshots wandered off to be alone with the giants. Others huddled together, faces illuminated by cigarettes. Drogo busied himself tinkering with his saw. Axel just sat, eyes closed, chin tilted upward, soaking in the sounds of the forest at night. I walked to the edge of our camp and lay on the ground. The sky was latticed with sequoia boughs, stars hard behind.

IN THE FOLLOWING DAYS, AWAKENING AMID THE SEQUOIAS BECAME AN especially keen delight. Birdsong woke me before the other hotshots did, robins and bluebirds warbling from hidden perches. When I opened my eyes, cobwebs bejeweled with dewdrops were strung between the trees, glistening in slanted bars of golden light. The sky was streaked with pink. The pine needles made a perfectly soft bed where I lay, warm in my sleeping bag, watching my breath steam in the crisp autumnal air until the others began to wake.

I felt at peace for the first time in months. I brewed a fresh batch of coffee for the crew before joining them for breakfast. To Aoki's chagrin, Axel had leveraged a personal connection at Patagonia, the retailing company, which was headquartered near our station, to secure nutritious dehydrated backpacking meals for the rest of our season. Instead of beginning the day with stale MREs, we enjoyed warm biscuits and gravy before our morning briefings. Maybe it was the food, or the air, or the trees, but in those mornings, I loved this group of men more deeply than I ever had before.

Had we not been working to protect the trees, it would have been easy to forget about the black column of smoke that grew closer each day. We fanned out in the forest and walked in a grid, searching for sequoias so we could scrape circles of dirt around their trunks to prevent flames from entering their bases and eating them from the inside. We cut down ladder fuels—mostly fir trees that had grown in the absence of fire—that would carry flames into the sequoia canopy. We set up sprinklers around the biggest sequoias at our campsite. Other firefighters arrived, bringing foil blankets, which they wrapped around the trunks to help repel the approaching heat. These actions felt small, but the care with which we performed them was a welcome break from the military aggression with which we were typically required to treat the world around us.

Others on my crew saw things differently. "These trees are two thousand years old," Axel mused to me. "They've seen plenty of fires. I'm sure they can handle another one." For once, Axel's views aligned with Red's, who told me that fires were natural, sequoias needed them, and we had nothing to worry about.

This sentiment trickled through the crew until everyone was joking about our duties even as we fulfilled them. Sequoias were invulnerable, the thinking went; we were arrogant for thinking we could help them. This dissonance prevailed until Márlon strolled up to the group with a sheaf of papers, a memo sent to us from the Forest Service. "Sequoias are adapted to low-intensity fire," he read aloud, "but extreme fire behavior poses a significant risk to their survival." In 2020, the Castle Fire killed an estimated 10 to 14 percent of California's sequoia population.

Axel blinked. He was accustomed to being the one who was up to date with the science.

"Oh. Really?" he said. "Damn. I didn't know that."

I was confused. Even though this crew had fought the Castle Fire the year before and had been closer to the sequoias' deaths than anyone, many hadn't grasped the scale of destruction as it unfolded around them. Somehow, amid the firestorm, the gravity of the event had escaped their notice. But throughout my days among the trees, as I scraped countless little lines of dirt around their trunks, the sentiment began to make sense. The idea that a fire could kill these trees stretched my own imagination. Sequoias, in their grandeur, are easy symbols of immortality.

This has not always been the case. From the moment of their discovery in the West, many Americans used sequoias as omens that could portend the future of our nation. Their discovery helped alleviate the insecurities of a young country whose citizens lacked a claim to their own history and whose culture was under perpetual negotiation. Europe had castles, cathedrals, and old cobbled streets, but the United States had the sequoias. The history of this nation, the thinking went, was seeded by God himself. Less than a century after its founding, these trees provided the United States with narrative roots.

Before long, threats to sequoias were seen as threats to the American self-image as an anointed nation destined for perpetual prosperity. Enemies of the United States recognized this. The year after the sequoias' discovery, while settlers were busy cutting the trees down, an English aristocrat commissioned botanists to collect sequoia cones to bring to the British Isles. Within the year, these botanists published a dispatch in the newspapers of London, announcing that so grand a species should be named for no lesser a hero than the Duke of Wellington, the great English general. This roused the American public to a fury. Bookish American botanists were irritated enough to take up politics, denouncing the English for scientific piracy. American newspapers blustered that the species should be named for General George Washington instead.

The English had the upper hand. Within only a few years, it appeared that the indiscriminate rapaciousness of American capitalism would entirely erase sequoias from their native continent. Meanwhile, the seeds took quickly on English soil. (There are currently more giant sequoias alive in the United Kingdom than in California.) What would it herald for the young American republic if we destroyed one of our only endemic claims to culture, transferring it in both name and habitat to the monarchy from whom we had so recently liberated ourselves?

This insecurity, projected onto the trees, was one factor that enabled politicians to eventually regulate lumber barons at a time when corporate political corruption ran amok. Once sufficiently protected, and with their scientific name wrestled from the grip of the English (a conflict that was not resolved until 1939), the sequoias resumed their mystique of endurance, becoming again quiet symbols of national perpetuity. This symbol, by sheer inertia, somehow continues to distort reality even as the sequoias burn.

FOR ONE SHORT CENTURY, THE SURVIVAL OF THE SEQUOIAS SEEMED guaranteed. Their natural range, 15 miles wide and 250 miles long through the Sierra Nevada mountains, was protected from corporations. And if corporations couldn't fell sequoias into oblivion, it seemed that nothing could.

Or, at least, this appeared to be the case as recently as a decade ago. By 2012, bark beetles were in the process of ravaging a record number of trees in the American West. Winters no longer froze deep enough to purge beetle populations, so the insects were untethered from past ecological limits. In California alone, in the years that followed, the beetles killed over one hundred million trees, turning whole landscapes into veritable firebombs. Yet even amid ecological collapse, the sequoias stood strong. *National Geographic* published an article in 2012 that assured the public they needn't worry about the sequoias. They were resilient. Their bark was red because of tannins they produced to repel bark beetles. They were dynamic. During wet periods, sequoias could drink up to eight hundred gallons of water per day. During droughts, they could close their stomata to conserve water. And they were unified. Their fused roots allowed them to hold one another upright through the most intense winds and storms of the ages. Just a decade ago, no one had ever documented a case of climate-caused sequoia mortality.

Not even wildfires were cause for worry. Of all the species on earth, sequoias are among the most fire-evolved. They are equipped to survive fire because they rely on fire. When smoke washes through the forest, the sequoias drop their cones. The cones land in the ash as it still smolders. The heat cracks the cones, dispersing hundreds of thousands of seeds. Historically, fires also cleared competitors and released nutrients in the soil, helping carry the sequoias between generations. Ever since the birth of this species, sequoias and flames coexisted in a symbiotic, harmonious relationship. Scientists once believed that sequoias, of all species, would be equipped to thrive in the flames of our new planet.

But then, in 2014, just two years after *National Geographic*'s reassuring report, something changed. The sequoias began to die. Like most climate calamities, their deaths couldn't be tied to a single clear cause. Rather, stresses accumulated, piling atop one another until they began to cascade. In the past, the trees had relied on a deep snowpack to provide a steady stream of water throughout the summer. But climate change exacerbated drought and produced warmer winters, causing less precipitation to fall, and most precipitation that did fall did so as rain. Without snowmelt, the soil was dry by summer. In vain, the trees tried to draw moisture from land that held none, sucking and pulling with such force that their internal water columns began to rupture from stress. Imagine the sensation of trying to breathe with a plastic bag over your head. For the sequoias, climate change was the plastic bag.

To alleviate this stress, the trees began shedding their needles. With fewer needles, the sequoias slowed their photosynthesis, reducing their need for water and allowing them to enter a state similar to dormancy. They were effectively waiting out the drought, like a human consciously slowing their heartbeat to reduce oxygen consumption until another breath can be drawn. But needles are essential for photosynthesis, which provides trees with the energy they need to survive. Without sufficient needles, the trees lacked the strength to ward off pests. Bark beetles, perhaps sensing this vulnerability, swarmed the weakened sequoias.

In normal conditions, when bark beetles attack a forest, the trees respond as a unit. The first victim releases chemical signals through its roots. Because the roots of sequoia groves are often grafted together, these signals can be transmitted to neighboring trees. In response, the neighboring trees bolster their defenses by increasing the production of toxins in the bark to deter the beetles. They may also share resources such as nutrients and water to the attacked trees, offering collective support. The attacked trees, perhaps sensing their own imminent demise, might reallocate their resources to healthier neighbors to aid their survival. Under normal circumstances, a grove of giant sequoias would be formidable. But climate change has disrupted these interactions. With reduced needle mass, the trees lack the energy needed for effective communication and defense.

Like the settlers of a century past, the beetles bored through the thick sequoia bark, laying eggs within. The eggs hatched into larvae. The larvae consumed the tree's phloem, the layer that transfers nutrients throughout their bodies, girdling them just as the settlers had. When scientists arrived to inspect the trees, they found them bleeding pitch in a desperate attempt to expel the insects from their bodies. The sequoias were being strangled, eaten alive; first one, then another, then a dozen, and a dozen more.

All of this was a preamble to the wildfires. In 2020, just a year before I entered that grove with the hotshots, a fire swept through these mountains. Trees that had survived hundreds of fires over the course of thousands of years torched like matchsticks. They died by the hundreds, then the thousands, then ten thousand, then more. The heat was so intense that limbs the size of elephants cracked and fell, plunging with their cones into a furnace. Afterward, when scientists searched the ash for survivors, there was nothing left. "We have never seen that before," said Christy Brigham, the head of resource management for Sequoia National Park, in an interview with *National Geographic*. "Never."

That single fire killed as many sequoias as all those killed by logging corporations during the years of California's genocide. Put another way, the fossil fuel industry is becoming the most destructive force these sequoias have experienced in their two hundred million years of existence—more destructive than the asteroid that drove a planetary extinction.

The night before the megafire struck our ridge, I took a final restless walk through the grove. I passed through the trees still living and among those that had died. The bare branches of the dead appeared as thin fingers spread against the sky. I wondered if there was still life coursing through their bodies. Not much time had passed since they burned. Sequoias do not die instantly. When they are cut off from the rest of the forest, they take years to wither away. I wondered if they could tell us anything. While the sequoias of the past have impressed themselves on the human psyche for their apparent immortality, those of today are equally powerful both for their unlikely survival and for the precarity of their continued existence. What do the sequoias portend for the delicate balance weaving together the rest of life on earth? What do they portend for us?

Maybe, in the era of climate change, the sequoias have emerged again as such poignant symbols due to a deeper intuition, or fear, that if the lines of our society—from the legal lines of regulation to the physical lines of dirt—fail to hold back the destructive forces driving sequoias toward extinction, then these lines might always have been but a diversion from a deeper, much more difficult kind of reckoning, one that is imminent, the far side of which is opaque and possibly terrible. Whatever it was, it was becoming visible now as a molten glow in the dusk, just over the horizon.

My head was spinning. In the darkness of that final night, the peace I had felt in the morning light crumpled into a trepidation that prevented me from sleeping. The stars were growing dim. The megafire was close. The next day, we were supposed to stop it.

⁂

IN THE FINAL HOURS BEFORE THE MEGAFIRE HIT OUR LINE, I WAITED ON the ridge with Barba. The forest was ghostlike in the shifting smoke. Our crew was scattered, finishing our final preparations. Axel and Scheer were downhill cutting a few more trees. Drogo was nearby slicing off any low branches he found to keep the fire out of the canopy. Aoki was everywhere, we assumed, a spirit in the smoke, measuring forces we could not understand before he decided on our next move.

Barba and I rested, saving our energy for the battle ahead. We sprawled in the dirt. My temples pounded from the smoke, my joints hurt, and I had worn holes through my leather gloves. But I was restless. For the fifth time that day, I pricked my thumb on my chainsaw teeth in a ritual that had become a nervous tic.

The megafire prowled nearby, somewhere in the gloom. Its proximity seemed to make everyone edgy, casting our personalities into sharp relief. The last I'd seen of Red, he had been sullen, spitting incessantly, deep in thought. The younger guys had become goofy. In a particularly surreal moment, I saw one imitating a TikTok dance through the smoke. Barba, for his part, just tore into his sack lunch, which another crew had brought up from base camp.

"Want your olives, Thomas?"

"You can have them." I wasn't hungry. The blend of exhaustion and adrenaline made my food taste like sawdust. Not so for Barba. I was frequently amused that I had been so intimidated by this man when I first joined the crew. Beneath the military tattoos, he was kind to his core. Once, when I complimented a hat he was wearing, Barba put the hat on my head and told me to keep it, then gave me his sweater as well, feigning anger when I tried to refuse.

"Crazy how your taste buds change, right?" Barba said now, on the ridge, wolfing down my food. "Like, things you hate as a kid, you love as an adult."

"Yeah. I eat everything now."

"Dude, same. Except mac 'n' cheese. I don't do mac 'n' cheese."

In the moment, it didn't strike me as odd that we were discussing mac 'n' cheese. Maybe it was better not to dwell on the approaching force or our own fragility on this line of dirt. We had done everything we could. Now, it took courage to construct a semblance of normalcy.

We sat together in the smoke, chatting like friends at a café, Barba telling me about his childhood. How when his parents brought him to Santa Barbara from Guadalajara, Mexico, at the age of ten, he didn't speak a word of English. How as a child, he learned to navigate this new place while his parents both worked sixteen-hour days. How he got sick after eating mac 'n' cheese while his parents were at work and had never eaten it again. Barba joined the Marines after high school to contribute to a country he felt had given him so much. But ever since those first days, he wanted to be a firefighter. "If I won the lottery," he told me, "I'd still be doing this. I'd be right here. I'd do this for the rest of my life."

We were interrupted by a figure in the smoke, huffing toward us. Barba and I stood, worried Aoki had caught us sitting. I relaxed when I recognized the burly shape of Jack.

Jack was hiking down the line to tell everyone to prepare for a nocturnal burn operation. At the wildfire's current rate of spread, it was arriving sooner than expected. We had completed the fireline that day, but we weren't yet prepared for the burn operation. To stop the megafire, we would need to light approximately ten thousand acres ablaze. In an ideal scenario, we would have multiple crews, helicopters, and airplanes to support us in an operation of this scale. We would have safety zones where we could flee if the megafire defeated us. At the moment, we had none of these.

Márlon's voice came through the radio. He was reporting from his lookout position with his characteristic cool. He estimated that we had about an hour left. He warned that the approaching flames were about three hundred feet high—a size most hotshots had only seen in training videos. He reiterated that we had about an hour before the fire hit our line.

Barba, Jack, and the others looked at one another for a moment, then resumed their chatter. Someone asked Jack about his ex-girlfriend. The man who inquired was less interested in Jack's well-being than in probing whether or not he was distracted enough to pose a hazard to us. Jack must have sensed this because he turned the inquiry into a joke, asking which ex we were referring to. The one who used scissors to cut his entire wardrobe into ribbons when he went to work? Or the one who currently worked as an escort in Tijuana? The laughter that followed was strained.

Márlon's voice came through again. He saw some smoke downslope, just below our line. It was a spot fire. The smoke was still white, which meant it hadn't climbed into the trees yet. We still had a chance to contain it. If we missed that chance, the fire would rush uphill and cross our line.

Axel replied on the radio. He was on it. A few minutes later, we heard him hiking up the fireline through the smoke. He told us to come with him. He clarified that if the spot fire got away from us and crossed our line, we *did* have a safety zone, but it wasn't really safe. It was a meadow where we would have a decent chance of survival. "We're calling it the Alamo," he said. He wasn't joking. All the surfer chill had melted from his demeanor. His eyes and movements were cutting. There was a fiery glow behind him. I watched a strip of smoldering bark twirl through the air and land on the fireline near his boot. He scuffed it out. "Let's go."

We left the fireline and hiked in single file across a narrow mountain spine where a few scraggly pines wrapped their roots around the rocks. A whistle brought our attention above. Márlon was crouched on an egg-shaped boulder. He pointed to his eyes, then down to the dark forest. From our vantage, we couldn't see the megafire, but we could see a few tendrils of smoke wisping up from the trees about a hundred feet below. The slope was steep enough that, if I were hiking with Kenzie, she would probably grab the back of my shirt and tell me to get away from the edge. We dropped down one by one to avoid injuring one another with rockslides, digging our heels into the scree and loose earth.

The spot fire had established itself in an open, rocky glade. The fire was chewing through grass and shrubs. A corner of my mind observed that we were lucky the embers hadn't landed in the fir trees around the perimeter. We split into two groups and attacked from different directions. Within a quarter hour the groups met in a boulder field. The spot fire was wrapped in a line of dirt.

Axel walked our line to ensure it would keep the fire contained. This allowed us a few minutes to rest amid the rocks. I briefly allowed myself to appreciate how perfect this stretch of land was; or, at least, how perfect it seemed to me. It reminded me of a place where Kenzie and I once camped about a mile away. We had pitched our tent at the edge of the forest. Behind us, the pawprints of a mother bear and her cub had disappeared into the darkness of the understory. Before us, we had watched clouds tumble down from the peaks. The clouds galloped in ghostly shapes, bending the grass and wildflowers of the watershed. I was so caught in this memory that for a moment I appreciated the smoke beginning to crawl toward my crew from the belly of the forest.

I wished I were back there with Kenzie. I allowed my mind to rest there as I dropped to my knees and clenched the saw between my thighs and pulled a file from my chaps to sharpen the teeth. I had nicked a rock while we were wrapping the spot fire. Even now, I was terrified someone would notice and make a spectacle of my incompetence.

Before I began sharpening the saw, I removed my earplugs. The sudden cascade of sensation brought me back to the present moment. I heard something like rolling thunder. It vibrated the air so that it seemed to come from everywhere at once. But unlike thunder, this sound didn't wane. It emanated from the trees. The forest was growling.

I tucked the file back into my chaps. I couldn't sharpen the saw.

My hands were shaking. Every instinct in my body told me to flee. The other hotshots looked like they felt the same. Everyone was watching Axel. It was a testament to how much we trusted him that we weren't already sprinting away. His eyes looked wild, but not afraid. He gave us a grin. "Let's get out of here."

We climbed back up to the ridge. Behind us, the sun was sinking into the fire. The colors of the western horizon were indistinguishable from those of incineration. To the east, the moon was rising.

While our group had been down in the glade containing the spot fire, other crew members had positioned our drip torches on the fireline. The torches were ready to go. We were ready to burn. Still, we waited for darkness to fall. Weather conditions would be more favorable then. With such a small group, we needed every advantage. I could feel the temperature drop. The humidity was rising. This would add moisture to the vegetation, slowing the intensity of the burn. Our tactical advantage would increase with the night. Some of the hotshots donned down-feather coats. The rest of us shivered.

When everything had slipped into an inky darkness, Aoki called for us to retreat back to the buggies. I was confused. No one else seemed to care. We hefted the drip torches and began our forty-five minute trudge back down the fireline. The smoke was lifting, but our footsteps kicked up a cloud of dust. I felt like we were abandoning this forest. I felt like we were turning our back on the sequoias and the world they represent. Just before reaching our buggies, I caught a glimpse of the megafire. Through a gap in the forest, I saw a lone flame flickering above the rest. Only a sequoia could carry flames so high.

But Aoki wasn't ready to abandon them yet. Before we drove out from the mountains, we stopped in the sequoia grove that had sheltered us the past nights. With our torches, we lit small fires around the trees, then left the flames to lick against the trunks and spread slowly through the grove. It was a small gesture, but it was all we had to offer, a final gift of nourishing flames before the megafire swept through.

⌘

FOR AOKI, OUR WITHDRAWAL WAS NOT A DEFEAT, BUT A DEFERRAL based on cold calculation. We left that night because he had noticed that the humidity was rising and the atmosphere was thickening, forcing the megafire to temporarily sink back into itself. If the megafire slowed enough overnight, we could return in the morning with more people. It was a gamble.

Aoki gathered us in a parking lot before dawn. "Today will be dangerous," he warned. My skin prickled. He had never said anything like this to us before. "Everyone up there will be looking toward you. You're hotshots. If you panic, everyone panics. So stay calm. Let's get this job done."

We rolled back into the hills at the head of an army. Three other crews were assisting us, with bulldozers, masticators, Black Hawk helicopters, and retrofitted commercial air tankers awaiting our command. We had all the technology of the twenty-first century at our disposal. This had become the highest priority wildfire in the United States.

Just before we lost cell service, I checked my phone for a final time. Even Kevin McCarthy, then the Republican Speaker of the House of Representatives, was paying attention. This was his district. The effluence of the megafire was visible from his home in Bakersfield. Much of Kern County was covered in ash. "We are racing against the clock," McCarthy later announced to Congress. "We cannot be the generation that allows these massive and ancient natural wonders to perish on our watch." As he said these words, his district was in the process of permitting the construction of approximately forty thousand new oil wells.

The hypocrisy was palpable, as expected. Vince Fong, a state assemblyman in McCarthy's district, feigned victimhood when scientists and public officials took steps to limit fossil fuel expansion. In classic doublespeak, Fong called the scientists and public officials opposing this extraction "out of touch" and "tone deaf," warning that limiting fossil fuel extraction "worsens everyone's quality of life" (Kern County already accounts for roughly three-quarters of California's oil and gas production and frequently suffers the most polluted air in the United States). Fong prefers forest management to decarbonization. Of course, no form of forest management will suffice if we fail to reduce carbon emissions on science-based timelines. However, politicians in the pocket of fossil fuel corporations have co-opted "forest management" as a buzzword that allows them to appear like they are doing something good, even as they prop up the industry at the root of the conflagrations. When I later asked the wife of a Kern County supervisor how much of the region's remaining oil reserves they planned to burn, her reaction was so vehement that spittle flew into my face. "We're going to burn it all," she said.

The hotshots and I arrived back on the ridge too late. Aoki had misjudged. Fires were burning behind us, ahead of us, all around us. The most we could do now was delay the megafire so airplanes could blanket the sequoias in pink retardant. We lit our drip torches and began trotting down the road, setting everything else ablaze. The winds were against us, so embers blew over our fireline. The supporting crews chased spot fires into the trees, trying to catch them before they could spread.

But the other crews were quickly overwhelmed. With Red, Márlon, Drogo, and several others, I piled into a buggy and took off to chase spot fires that were establishing themselves deep in the forest. The battle for the sequoias was lost. Now we just needed to hold off the flames long enough to keep them from blocking the road before everyone could escape. Driving down old logging roads so overgrown that branches stripped paint from our vehicle, we got lost. Sleep-deprived and under pressure, the driver snapped, cursing, and swerved the buggy into a three-point turn, almost flipping us—a potentially fatal mistake that would have stranded us in the path of the megafire. Márlon, calm as ever, ordered him to stand aside while he took the driver's seat. Drogo cut down a tree so we could turn around.

We worked through the night, splitting up, doing what we could to save the homes in the fire's path. Some of the hotshots jumped on UTVs with chainsaws and radios to hunt spot fires in the darkness, guided by a thermal drone. Scheer, defying orders to protect only the most expensive homes, led a bulldozer around a group of trailers because he knew that their inhabitants, like him, could not afford to rebuild. I worked with the others to start fires around the rest of the cabins. We guided the flames outward so that when the megafire arrived, it might jump over the homes.

I was overwhelmed by exhaustion and the unfolding destruction. Reality quivered. In air that tasted like pine, I hallucinated flurries of snow, the smell of cozy hearths, and flickering holiday lights. But the snow was the effluence of burning buildings, the smell was wildfire smoke, and the lights were indicators that the local power company hadn't shut off the electricity, which meant that the voltage could jump through the smoke and electrocute us.

The others were losing their senses too. I found a few younger hotshots crouched over a campfire they had built in the ashes to warm themselves. The night was bitter cold. "They won't know we were here," one of them muttered, looking at the cabins we had saved. "They'll see that the fire went right up to their homes, and they'll think they got lucky. They'll never know we were here."

"It's not real," another said. "It's not real." He looked at me with eyes that were black in the firelight. "None of it's real." He giggled, wrapping his arms around his knees and rocking himself in the cold dark.

Aoki was a point of solid ground. He busied himself with his maps, planning a new route of attack. He didn't appear fatigued or fazed. But when the horizon brightened with the dawn, his phone rang. As he listened, I saw something lift from his face. Maybe his expression was joy. Perhaps pride. He told us to gather around.

The phone call was about the sequoia grove that had sheltered us those many nights. The same grove where we had scraped those little lines, removed ladder fuels, and swaddled the trunks in blankets. Where we had made a final offering of fire before we fled.

When the megafire hit that grove, Aoki told us, it had dropped to the ground and crawled through the needles where we had slept. While the rest of the sequoias were incinerated, not a single tree in our grove had been killed.

Part V

REGENERATION

CHAPTER 13

THE FIRE SEASON FIZZLED OUT. THE YEAR HAD BEEN THE FOURTH hottest on record. Climate-driven disasters in the United States killed at least 688 people, inflicted some $145 billion in damage, and contributed to the extinction of multiple species. Fossil fuel companies enjoyed record profits. Over fifty-eight thousand wildfires burned seven million acres. We fought fifteen of those. We survived.

On the last day of fire season, we piled back into the classroom where our season had begun. Many scientists now consider the idea of a "fire season" to be a misnomer because climate change has produced year-round danger. A few years ago, the largest California wildfire ever recorded at the time had burned through January right there in Santa Barbara. But, for us, the fire season would officially end at noon that day, November 5.

The classroom felt different now. The leaves outside the windows were colored gold with autumn. The river was filling with water. Inside, the informational posters on the walls, once mysterious, had become as mundane as the alphabet. I knew the common factors of fatality fires, the environmental shifts that should make my hair stand on end, the different ways of cutting a tree. I also knew the names and stories and quirks and fears of everyone around me, and they knew mine. Even the squelch of tobacco sounded different now. Some of it came from me.

We waited as Aoki called us into his office, one by one, so he could fire us. By firing us, he was enabling us to claim unemployment during the offseason. If spent carefully, that meager trickle of unemployment money could tide us over until the fires ignited again. We had worked a year and a half worth of hours in just six months. Most hotshots would spend a month recovering their health before resuming a training regimen to prepare for the next fire season.

Just before noon, Aoki walked into the classroom. He stood at the podium, black hair falling down his back, his eyes dissecting us under thick brows, jaw long and set. His face was a bit puffy. A few days before, he had left the station for an emergency root canal. He refused medications so he could come back to the station immediately afterward. He showed no sign of pain.

I was expecting a speech—something rousing maybe, a celebration of our achievements or a commending of our improvement. But when Aoki spoke, he just sounded tired. I realized this wasn't the end for him. It was a bridge into the next season, a bridge he had already crossed many times. Aoki started by telling us he had only visited two restaurants since the pandemic began. Once, he walked down the beach to get pizza. And the other time, with us. The pandemic was awful, he said, but the solitude agreed with him. All of this was a preamble to say he had heard a rumor he was going to join us at the bars to celebrate. That rumor was wrong. He was going home. He would sit on the couch with his wife, listen to the waves, drink a glass of whiskey from the bottle we had bought for him, and pet his cats. If we wanted to see him, we could come to the station and hike with him.

Aoki called us to the front, one by one, releasing us. He shook our hands and gave us a gift. When it was my turn, he gave me a belt buckle. It was the same one he wore. The belt buckle was small. It fit in the palm of my hand. The metal was cold and shaped like a shield, inscribed with our crew's emblem. Most people would not know what this belt buckle meant. For those in the world of wildfire, the buckle would demand automatic respect. This, I had previously thought, was why the tradition of the belt buckle persisted. Now, looking around at the buckle on each of my crew members' belts, I realized that the buckle was not just a badge of honor; it was a way to identify a family I had not had before.

We gathered on the lawn outside the station, chatting and laughing. After all our anticipation of the end of the season, it was difficult to leave now. We stood in a crowd until a whistle sounded. Over the radio, we heard the same dreaded, squawking voice we had heard many times before, announcing a new ignition in Big Sur. Everyone stopped, smiles frozen on our faces. Through the window of the station, I saw Aoki's long form unfurl from his desk.

Red broke the tension. "Get out of here, you bastards!" he hollered, flapping his arms to herd us off the property. "Get through the gate! Go! Go!" The moment we stepped off the property, we were free. "Run!" Red yelled after us. "Run! Run!"

Twenty young men ran toward their trucks, toward the shade of the oak trees. The season ended like it had begun. We peeled off our shirts, opened the tailgates, cracked beers, and started getting drunk.

THE LAST TIME I SAW JACK WAS THE DAY AFTER THE CREW PARTY. IT was two in the afternoon. After a big breakfast with a mimosa, I was feeling slightly recovered from the previous evening when Jack called me for a ride. He told me he had met some people at a club and went to an after-party. Twenty-six hours after we had been released from the station, the after-party was still in full swing. Jack was crashing, so I drove to pick him up. I found him at a shack in the foothills.

It was strange to see Jack in his civilian attire. He was wearing checkered Vans, tight jeans, and a hipster hoodie—the same as the night before. His beard was ruffled and his eyes were red, but he was composed. I knew where his truck was parked because he had led me there the night before. He had said he wanted to show me something. We had jumped a fence we could have walked around and ran down railroad tracks and jumped another fence to find the parking lot. The surprise Jack wanted to show me was a homemade bomb. He called it a firework, and he detonated it at an empty intersection. I ran away from Jack then, hearing sirens approach. Now, I pulled back into that parking lot in the light of day, found his big red truck, and asked if he was okay to drive. He obviously wasn't, so I turned around and drove him back over the mountain pass to his trailer beside the hotshot station. When we hugged goodbye, his eyes were bright with emotion. He thanked me for the help.

After I gave Jack that ride, he began calling me every Sunday as he returned from San Diego, where he spent weekends with his daughter. On the third week of this ritual, I was in my pajamas preparing for an early retreat to bed. Jack wanted to come by my apartment with a bottle of whiskey. I gracefully declined, prompting him to admit that he felt like he was in my debt. "One day I will no longer owe you a favor, Sir Thomas," he barked into the phone. After he hung up, I decided someday it might be useful to be able to call in a favor from Jack. It all felt like a game. Then one day, he stopped calling.

After two consecutive weeks of silence from Jack, I decided to check in. He didn't answer. Another week passed. I kept calling.

Then, while I was cycling into town for yin yoga, flanked by a red sunset ocean and purple mountains, my phone rang. It was Jack. "Shit's been going crazy," he said.

Jack told me that his ex-girlfriend had COVID, so Jack took their daughter. His ex called the police, accusing him of kidnapping. But he thought it would be negligent to expose his daughter to COVID. His lawyer agreed, so—

"Hold on," I interrupted. "You have a lawyer?"

"Yeah, dude. I'm forking out one or two grand a week to her. Let me back up."

Jack was in the midst of a custody battle for his daughter. He had hired the best lawyer he could find. It cost him sixty dollars every time the lawyer opened an email from him. All his money from the season was already gone, but he felt like he was winning the fight. They had gone to court. The judge agreed that Jack should have custody of their daughter, but his ex said Jack couldn't have custody because he was a hotshot. The judge agreed: a hotshot couldn't raise a child.

"What?" I had stopped bicycling, abandoning my yoga plans. "Seriously? That's messed up, Jack. I'm so sorry."

"No, dude. It's absolutely correct. It's a completely valid point. Hotshots can't raise children alone. I can't raise a daughter if I'm gone six months out of the year."

"Jack."

"Yeah, man. So it sucks, but I'm not coming back next year. I'm transferring down to San Diego."

"What are you going to do?"

"I'm getting on an engine on the National Forest down here. I'm going to raise my daughter."

"Man."

"It sucks. I don't want to be on an engine. I hate engines. You get fat. Everyone makes fun of you."

"Just wear your belt buckle, Jack. No one will make fun of you." I didn't know what else to say. The family I had found on the crew was already starting to disintegrate, just weeks into the offseason. I didn't want Jack to go.

I was standing there in the night, and Jack was still talking. He didn't want to live in San Diego. He had been born there, but it was a ruined city now, overrun with liberals like me. But he'd suffer the culture for his daughter.

"If you don't like San Diego," I advised, "you could sell the house and build your own compound in the Mojave. Bring your guns, your dune buggies. I'd visit."

"Valid point." A spark of humor crept back into his voice. "That's a great idea, Thomas."

IN SANTA BARBARA, ON A RAINY DECEMBER NIGHT, WHEN THE FIRES were all out and the hotshots unemployed, Scheer burst into my apartment. I was alone; Kenzie was out kizomba dancing. The rain had washed out the mountain road to Scheer's trailer, so he was searching for a place to stay. He was drenched. His hair was disheveled, one eye bruised, and several beer bottles clinked in his arms. Through the open door, I heard the crash of ocean waves.

"Whoa." He paused in the doorway. "You've got a vibe going. What've you been doing?"

I had been meditating. I just told Scheer I was cleaning.

"You've, like, cleansed this place, dude. Here." Scheer took my phone and put on Tool. "This is the level I've been on." He bobbed his head to the heavy rhythms and howling refrains. He had spent the entire day at a jiujitsu gym. He didn't know jiujitsu, so he had been severely beaten, but he refused to quit, even when they choked him. The gym members thought he was weird.

Scheer sat at my table and opened a beer. I sat on the couch. We were both still emaciated from fire season, with cavernous cheeks and hunched shoulders. Firefighters call this the hotshot body. Scheer let out an exaggerated sigh. "So," he asked, "like, what do you *do* at home? I mean, I know all these people who sit at home. What do you do? Just sit here like this?" He twiddled his thumbs.

I couldn't think of an answer. I had never considered that it could feel strange to be home. "I guess I work in the morning," I finally said. "I write a lot. Read books. Usually exercise in the afternoons."

"Do you watch TV? No, you don't have a TV. I just don't get it."

"Well, what do you do? Just wake up and go surfing?"

"Yeah. I went and got my ass kicked at six thirty this morning. Or I drive up and down the coast looking for waves."

Everyone, I knew, had their own ways of rebuilding their bodies and minds after fire season. I meditated and focused on research. Scheer surfed and fought. Red went fishing. Aoki went to Fiji with his wife. Most drank.

Scheer's phone buzzed. It was Barba. Scheer told him he caught me meditating. Barba thought that was funny. He invited us to a bar. I offered to drive.

In the car, on a wet road reflecting streetlights, Scheer was unusually silent. When he finally spoke, his voice was low.

"Isn't it weird that we'll be doing all that crazy stuff again in a few months? Like, everything will be chill, then suddenly we'll be in a forest and Aoki will be like, 'Burn it all! Save the cabins! *Go, go, go!*'"

"Yeah, man. Weird thought." Guilt settled in my stomach. I hadn't told Scheer that I would be unable to return to the crew.

"I think that's why I have such a hard time in the offseason," Scheer admitted. "It's just, you're going one hundred percent for so long; you're gone for six months, then everything drops down to zero. There's no stability. It's a headfuck going so hard, for so long, being surrounded by everyone. . . . You hate them, but you love them. Then you lose it all."

◈

SCHEER AND I ARRIVED AT THE BAR JUST AS BARBA DID. BARBA CLAPPED us on the back, gave us long hugs, and whispered in my ear that he had missed me. He told me he had traveled to Washington State to buy a fancy cat with his girlfriend. He sent Aoki a picture of the cat sitting on his head. "Aoki's a crazy cat lady," Barba announced. "El jefe loco has like three cats, bro!"

Barba made his rounds through the bar—he seemed to know everyone there. He told me that, besides the trip to Washington, he hadn't left his house in the three weeks since our season ended. He had binge-watched every season of *Peaky Blinders* with his girlfriend. Then he had gotten a call for an interview with a city fire department.

That stopped me. "Wait," I asked, "so you're done with the crew?"

He was. He wanted to build a life with his girlfriend, but he knew he couldn't step away from fire. So he was going to work for the city. It had been his dream since he moved to America as a child. He would be home more. He would make more money. But he wouldn't be on our hotshot crew. That last bit went unsaid, another unspoken hole forming in our family.

A shape filled the doorframe, interrupting us. Drogo entered the bar. I had forgotten how tall he was. He was dressed sharp—ironed slacks, a button-down shirt, cotton coat, and stylish man boots. With his square, shaven jaw and tight haircut he looked like he belonged more on Wall Street than the fireline. Barba disappeared into his bear hug.

When Barba squirmed free, Drogo asked what he'd been doing for money. "Same as you, homie," Barba said. "Welfare queen."

"Know of any under-the-table jobs?"

"Why? Cause I'm Mexican?"

"Well, yeah."

"Landscaping."

"What about construction?"

"Do construction."

"Can I get paid under the table?"

"Hell, yeah! What, unemployment ain't good enough for you?"

"I want more," Drogo replied, downing his beer. "I always want more."

"Well, tell you what. Go stand outside Home Depot with the rest of my people. Get picked up. The boss will see your white ass and choose you immediately."

"Yeah, I was thinking I could exploit the closet racism of the old white men doing the hiring."

I had been listening quietly to the banter, but this was too much. "Are you serious, Drogo? You can't just go steal people's jobs!"

Barba laughed and punched me lightly on the shoulder. "Ay, Thomas, easy with the savior complex. My people don't need you."

Rain was washing down the windows and pattering the roof. Our voices were as comfortable as the warmth of the bar. We were grasping for something, a world we had been so excited to escape. I found myself alone at a table with Drogo. He was drumming his fingers, clenching his jaw, mischievous eyes following the antics of a rowdy drunk. Drogo told me he had just returned from a bender in Las Vegas.

"Whoa," I said. "So that's your life right now?"

"No." He shook his head. "That was an exception." His face was calm, but something was boiling behind his eyes. "This is my regular day. I wake up every morning. I do a pour-over coffee. Good coffee, and I just pour the water over it. Very zen. You would love it. Then I take the coffee and walk to the beach in my pajama pants. Then I go back to my place, and I think about blowing my skull off." Drogo smiled, all teeth.

I tried not to dwell on that comment or on the suicide rates of hotshots. The alienation was acute for everyone, accompanied by flashbacks that spiked adrenaline, anxiety, and depression. I required both earplugs and noise-canceling headphones whenever I was around landscaping crews. The sound of two-stroke engines made blood rush to my head and my heart tremor. And I was better off than most. Recent research pegged our suicidal tendencies at around double the national average. There was no offseason support. Hotshots charted their own routes to recovery.

"Are you bored?" I asked Drogo.

"Too bored."

"I just got an ice ax. Let's do some mountaineering."

Drogo perked up. He suggested Mount Whitney, the highest mountain in the lower forty-eight states. I offered that maybe our first winter alpine expedition shouldn't be that one. He scoffed and listened skeptically as I suggested a more intermediate route behind Los Angeles, a ten-thousand-foot mountaineering trial run on a route where only a few people died each year. He agreed with a compromise.

"Let's do that one," he said. "Then we'll grab some grub, drive through the night, and mob up Whitney the next morning."

We shook on it.

DROGO AND I WERE LOST ON A MOUNTAIN IN A BLIZZARD. THE ALL-encompassing whiteout distorted any sense of up or down, giving me vertigo, like I was floating in a cloud. But it was a cold cloud, and we had been soaked to our skins from rain at lower elevations. Now, our clothes were hardening to ice. But Drogo was loving it, and that was what I wanted to see.

We had just begun our final ascent—a half-mile stretch of near vertical snowpack that required our ice axes—when Drogo told me he was getting worried about frostbite. I wasn't surprised. That morning, I had found him at his Malibu studio apartment shirtless, barefoot, and wielding a hatchet. "To climb the ice," he explained in response to my puzzled expression. I provided him with a real ice ax I had borrowed from my neighbor—mostly so he wouldn't frighten the day hikers—but I didn't have gloves or a coat that fit him. He compensated with a leather jacket and tube socks on his hands. Those socks were now frozen stiff, and he couldn't feel his hands.

Nevertheless, Drogo was laughing, plodding along right on my heels. "I miss this shit, man!" he kept calling over the wind. I did too. Maybe we had become addicted to a physical grind, or at least to the euphoric loss of self that occurred when we pushed past a certain threshold of pain. Or maybe it was the bond we missed, the unspoken friendship that, for many men averse to vulnerability, can only be forged in the shared experience of discomfort.

I was having a great time too, despite the cold and the howl of the wind. I was back in the immediacy of the moment, swinging my ax, pulling myself up, kicking my crampons in the snow, catching my breath, and doing it again. I was also relieved that I had convinced Drogo to climb the smaller of the two mountains. I learned, at the trailhead, that Drogo had nearly died summiting Whitney the year prior, the day after he was released from the hotshots. He had passed out from the cold on the summit and woken with frozen joints. And there had been no snow.

Drogo called for my attention. He was grinning, eyebrows frosted and snot frozen on his nose, but he told me he was turning back. He didn't want to lose any fingers and no longer trusted his hands to grip the ice ax. I offered to turn back with him, but he urged me on. We would meet at his truck.

I fist-bumped him and continued, picking up the pace, aware that the blizzard light was dimming as night fell. I pulled myself up and over a pile of boulders and climbed higher, toward the summit. It was close.

The clouds parted for a moment. I was secure with my ice ax and crampons, but I was standing straight, and my chest was pressed against the snow, which meant the slope was almost vertical. I looked down. I shouldn't have. The world seemed to spin beneath my feet, a monochrome spiral of white and black. The black shapes were rocks. They looked jagged. For a moment I reeled, feeling I could free-fall onto them.

I pinched my glove in my teeth to free one hand so I could check my map on my phone. I was only a quarter mile from the summit. I could see where this slope leveled out onto a hill with a tree at its top. I felt like I could touch it. I was so close. But I was also calculating, using the sort of on-the-fly metrics I had learned on the fireline. I could get to the top. But I would descend in the dark. I had a headlamp, but it would be useless in the blizzard. And to get down, I would need to glissade, a tricky practice where you sled on your rear, using the ice ax as a rudder. My only experience with glissading was a ten-minute YouTube video I had watched that morning, so I wasn't confident I could avoid the rocks. If I broke a leg in the dark on my descent, no rescue team would be able to find me in the blizzard. There was a nontrivial chance I could die. After surviving the fire season, that would be a terrible embarrassment. I stared at the summit for a moment longer, swallowed my desire, and turned back.

This was a good decision. I reached the bottom quickly because I lost control of my glissade and tumbled headlong over rocks. I checked my bones. None were broken, but I had been screaming as I fell, so I felt sheepish when I found Drogo. He was shivering under the awning of a boarded-up ski cabin, waiting to make sure I made it down. We were too cold to speak, so we started walking down the trail until the snow turned to rain and the ice to mud.

By the time we reached Drogo's truck, our shivering had intensified to the convulsions of early onset hypothermia. At first, our hands were too numb to crack beers. We cranked the heater and sat, groaning as our thawing nerves sent needles of pain into our hands and feet.

"Still want to drive north and mob up Whitney?" I asked.

"Let's go," he said, putting the truck into gear.

"You're joking."

"Let's go get a burrito."

The road plummeted down from the world of waterfalls and snow and pine forests into a field of light that stretched as far as the eye could see to the north and south, ending abruptly in a black pit of ocean to the west. We rehashed our climb as we had a dozen fires, the pain forgotten, replaced by small moments we could laugh about. I asked if he had heard me screaming when I tumbled over the rocks. He hadn't, but he admitted he had urinated on his hands to stave off frostbite while he waited for me. I told him I didn't think that was effective. He shrugged. He had seen it on television.

Our conversation turned to the crew, and to who we had lost. Edgar was gone. After twenty years, he had retired from the hotshots for a desk job. Edgar, Drogo told me, was the nicest person in the world off the fireline, a cuddly bear who would bend over backward to help us in our careers. Axel had retired as well, joining a city fire department so he could be with his kids as they grew up. Barba was gone, and Jack. Márlon would never leave. With a new baby in his house, he joked that the fireline was the only place he could find solid sleep. Scheer was coming back too. He would have an identity crisis without the hotshots.

"Next season is going to be epic," Drogo said. "You, me, and Scheer on the saws. We'll be running the crew."

If I was honest with myself, this was why I had spent the day hiking with Drogo. I couldn't think of a better way to tell him I was quitting the crew. I had thought that maybe risking our lives together would lessen the sting. Drogo was frightening. He was emotional. He was also one of my closest friends. But now, with a cityscape of concrete and neon lights blurring the world outside his truck, letting go of our bond was more difficult than I had imagined.

I told him. He didn't take it well.

"*What?*" he growled. I heard the leather of his steering wheel creak under his grip.

"I can't come back. I'm sorry."

"We spent the whole day together . . . talked about the crew all day . . . *now* you tell me?"

Maybe I had hoped Drogo would help me fill in the new void that had settled beneath my chest when I made that decision. Maybe I had hoped he would help me understand.

"I didn't want to ruin the hike," I offered lamely. "There's no good time to say something like that."

"So that's your game? You just get the belt buckle and bail?"

"It's not like that."

Drogo began taking deep breaths and exhaling violently. It was a technique he had read about to control his anger. I had seen him do the same as a puller when branches gouged his arms and trees cracked against his helmet. "What did Aoki say?"

When I had called the station to resign, Aoki's captain picked up. He told me I was dead to the crew. The words hurt. I couldn't imagine being cut off from the people I had risked my life with for the past six months. I had lost sleep over the decision. I had called my mom. This is what I had feared—that I had betrayed the crew, several of whom, including Drogo, had become my closest friends.

After a moment, in the background I heard Aoki laugh. "I'm just kidding, man," his captain said. "Thanks for all your hard work. Come by the station anytime."

Once the initial shock settled, Drogo calmed down. He pulled into a Mexican restaurant and bought us margaritas. He wanted to talk about my decision.

It had started with Kenzie. Aoki had been right. The fire season had stretched our relationship to the breaking point. For the first time, Kenzie told me, she had begun seriously imagining life without me. She admitted she had been looking at different places to live. The tension arose, she thought, from the fact that we had needed to build entirely different worlds apart from each other. She had mastered kizomba dancing and become an instructor. She began cooking vegan recipes every evening with seasonal produce. She leaned heavily into a new job organizing food sustainability events for Coachella, Sundance, and the United Nations. She had been excited for me to come home. She wrote lists of fun things we could do together to rebuild our connection. But when I arrived, I would crash. She said I was frequently traumatized. And she experienced whiplash because, after fully embracing her independence, she would suddenly need to shift into a mode of nurturing and care, like mothering, which was difficult when we had lost our emotional connection. When I was home, she spent the first two days patching me back together, and the third day preparing to send me off again.

On my end, to survive the crew, I had been forced to deaden some of my emotions. I had built walls around my fear, my sensitivity to pain, and my response to ridicule. I tucked everything away into a shell that no one and nothing could crack, where I could sit safely within myself. But emotions cannot be numbed selectively. By sealing off the world, I was sealing out Kenzie as well, and she could feel that. We each had done what we needed to survive, but a chasm had formed between us. We could see each other on the other side, but we didn't know how to cross over, and it was a cold and lonely space between.

I wasn't ready to give up on Kenzie. I wasn't ready to give up on the future. And when, a month into offseason, I received a phone call offering me a job to research what communities were doing to rehabilitate forests and return good fire to the land, I saw an opportunity to repair my relationship with both.

Drogo was listening quietly. Something had shifted between us already. We weren't colleagues anymore. We weren't a crew. But he was nodding, offering small words of advice. He had become an old friend.

"I'm proud of you," he said, when the margaritas were gone and we stood in the glow of streetlights to part ways.

"I'm proud of you," I countered.

"You're still a hotshot," he said, wrapping me in a hug.

I walked to my car and held on to my tears until I was on the highway. I took the long way home, following the mountains north. They were as black as the ocean under starlight. I could imagine how a wildfire would race up them, exploding the homes and terrifying the city. But I wanted to imagine something else. I knew there was a counterforce of healing at work, somewhere. I drove north in the darkness, looking for the light.

DURING ONE OF THE DARKEST MOMENTS OF THE FIRE SEASON, I BE-came enthralled with a deranged fantasy. I wanted to bring my dog to a beach in Big Sur where the elephant seals congregated. I wanted to watch her stare in wonder at those majestic creatures, whose musky smell and garbled bellows might unearth the primitive memories buried in her DNA.

This fantasy became a possibility in the afterglow of the fire season, when Kenzie talked me into joining her for a backpacking trip. Big Sur was one of the most scenic places on earth, we agreed, and I needed to heal my relationship with it. My associations had been rewritten as a ghoulish kaleidoscope of screaming chainsaws, screaming people, physical collapse, dark thoughts, dying land, burning redwoods, and incredible stress. Kenzie wanted to help me plant some seeds in the ash of my memories so something beautiful could grow between us again.

The night before we left, I almost canceled. I tried packing my backpack and needed to stop. I was too anxious. I felt like I was packing for a fire. I called Kenzie and told her I couldn't do it. Maybe we should go for a day hike and get dinner. "Whatever you need," she said. But I gritted my teeth and finished packing. For most of the night, I couldn't sleep. I felt like someone would kick me awake to charge toward a fire before sunrise.

I woke, instead, in a golden light, the sun already risen, a gurgle of coffee coming from the kitchen. Kenzie let me take my time. On the drive, she let me pull over to look at wildflowers. "You're not going to work," she kept reminding me.

We stopped at the gathering point of the elephant seals. We were alone. I let our dog out of the car, watching her intently. She sniffed the salty air, then cocked her head at the hoot and slap of the seals wrestling one another in the surf. I led her down a gravel path to the lookout. She sat, wind blowing locks of fur from her eyes, staring through a fence toward the raw spectacle of four-thousand-pound pinnipeds warbling strange songs.

Except she wasn't watching the elephant seals. Another tourist had approached with a dog, and that was what had captured her attention. They circled each other, noses jammed under each other's tails. The other dog nearly urinated on our dog's snout. Whatever our dog smelled in that urine neutralized any remaining interest in the seals. Then a horde of camera-wielding tourists arrived in a bus and marched toward us with flopping fanny packs. I gathered Kenzie and the dog and made a panicked escape, feeling moderately satisfied.

We reached the trailhead soon after. I pulled my gear from the Prius and prepared to hike—which took me all of thirty seconds—then realized that my heart was racing again. I needed to slow down. I asked Kenzie to lead so I didn't black out and race up the mountain at hotshot speed. This was a vacation, I reminded myself.

We climbed into the mountains, over flowery knolls where the ocean appeared as a sun-touched disk of blue and gold. Oak trees formed tunnels of shade over the trail. Most of the streams were dry. When we passed manzanita bushes, I imagined how I would cut them, how a fire would burst up the gullies, how it would feel to be cutting line up these slopes again. Fallen leaves crunched under our feet. Kenzie commented on what a nice autumn hike this was. I corrected her. These were evergreen coastal species and the leaves were falling because of a drought that was driven by climate change. Kenzie looked at me funny. I wanted to slap myself. I tried to see the landscape as she did, as I once did.

Late in the day, we ascended high enough that the oak trees gave way to conifers. We found the southernmost groves of redwood trees in California. The first one I saw was dead. A fire had killed it, I noticed, by entering its base and consuming it from the inside. The other redwood trees were alive, but they were surrounded by white firs that had colonized the shade and would carry flames into the redwoods' canopy, killing them if a wildfire burned through here. I pointed this out to Kenzie. She gently reminded me to enjoy myself and to stop looking at the forest like a sawyer.

We arrived at a rushing creek—a river, by California standards. It was a fairy place, with huge, moss-covered boulders, redwoods providing shade, ferns carpeting the ground, and a waterfall tumbling through it all. We were blissfully alone. I climbed into a pool at the base of the waterfall to rinse off my sweat, then dried myself on the bank against a log. Kenzie came to sit beside me.

"What do you see?" she asked, gazing into the pool.

"What do I see?"

"What do you see in the water?"

This made me smile. Her words carried me back to a conversation we had had on one of our first dates in Kansas, back when we were in college. We had hopped a barbed-wire fence, trespassing on ranching land so we could hike into the Flint Hills. There, we discovered a perfect stream of clear water. I had wanted to kiss her then, but I was too nervous. Instead, I asked her what she saw in the water. I wanted to know whether her eyes noticed the light glancing off the surface or the dark shapes in the depths.

Now, Kenzie returned the question to me. If I allowed my eyes to rest on the water, I could see a reflection of the forest and the sky, ripples of sunlight like crystal threads. But if I allowed my eyes to settle, the view shifted. The depths of the pool were opaque, with occasional glints of mineral coming from piles of smooth river stone. Saying nothing, I stared into the drifting water.

"Remember to see both," Kenzie told me on the bank, taking my hand in her own. "You can see both at the same time."

CHAPTER 14

IN THE MOUNTAINS ABOVE LOS ANGELES, I WAS CRAWLING ON MY stomach through a forest. Jeffrey pines towered above, filling the air with their distinctive aroma of sweet vanilla. Everything was silent except for the rattle of woodpeckers and the swish of wind in the canopy.

I had traveled to these mountains to help some Forest Service ecologists plan a burn. On August 12, 2020, during an onslaught of megafires, California governor Gavin Newsom had joined the chief of the U.S. Forest Service to sign a memorandum in support of prescribed fire. "The health and well-being of California communities and ecosystems depend on urgent and effective forest . . . stewardship," the memorandum read. "California's forests are naturally adapted to low-intensity fire . . . but Gold Rush–era clear-cutting followed by a wholesale policy of fire suppression resulted in the overly dense, ailing forests that dominate the landscape today."

This memorandum put a stamp of government approval on a

burgeoning scientific consensus: fire, when applied the right way, in the right doses, in the right ecosystems, could make forests more resilient to climate change and reduce wildfire hazards. To increase the land's exposure to healthy fire, the state and federal government pooled together tens of millions of dollars and poured it into projects. The money, they hoped, would result in the growth of prescribed burning. "We really don't need any more research to know what we need to do," Malcolm North, a professor of forest ecology at UC Davis and researcher for the U.S. Forest Service, told me. "It's pretty damn clear." We need to be igniting more of the land.

And yet, despite the science, money, and government support, prescribed fires are spreading slower than expected. California's goal is to treat one million acres annually by 2025. In 2021, California burned a mere sixty thousand acres. The next year saw little improvement.

Every attempt to mend California's relationship with fire seemed to sink under the weight of mistakes, public outcries, litigation, and bureaucracy. To learn how people plan prescribed burns, and why they fail, I joined Nicole Molinari, the lead Forest Service ecologist of Southern California, and her team in the mountains.

"We have two hundred one-hour fuels," I shouted to Ryan Fass, a mustachioed ecologist with a clipboard. Then I crawled back to begin counting thicker pieces of wood, adding to a growing data set that would help the Forest Service predict how a fire would behave in this ecosystem. There were only four of us, with much information to gather: we needed to count each species of tree (alive and dead), assess the percentage of ground covered by debris, and measure the thickness of soil layers to a tenth of a centimeter.

All that data, we hoped, would provide a glimpse into what might happen if the Forest Service carried out a prescribed burn here. When we left the forest, Nicole's assistant would compile our data and send it to a computer modeler, who would play with variables to simulate how this environment would burn in different conditions. Afterward, the modeler would send the information to the firefighters charged with conducting the burn. Eventually, if the environmental conditions aligned, if the paperwork was approved without a hitch, and if the surrounding communities maintained their support, Nicole's team would return the forest to its historical pattern of frequent, light fires. The goal was to protect the region from the kinds of large-scale conflagrations that can permanently erase forests. The on-the-ground reality involved counting thousands and thousands of sticks.

When we paused for lunch under a giant Jeffrey pine, I chatted with Nicole. Small and spritely, Nicole had the birdlike buoyancy of someone who can't drink coffee without their energy brimming over. She had a PhD in plant ecology and managed six thousand square miles of California's public land for the Forest Service. Not the academic slog, not the massive responsibility, not even the bureaucracy of the federal government seemed to have dampened her deep appreciation for each habitat.

"People like to call these sky islands," Nicole told me, gesturing to the forest around us. It was an apt metaphor. The mountain rose seven thousand feet from the coastal area, creating a coniferous oasis in the sky. Below, shrublands and cityscape stretched in every direction. That level of isolation renders plant life precarious. "When you get a high-severity fire, the whole forest is gone," she said. "If all the trees burn, there are no sources to reseed it."

Nicole and her team of ecologists expressed hope that science would point to the best course of action for conserving forests, and they were devoted to their cause. To gather data, they often camped for weeks in rugged terrain, rising with the sun to spend twelve hours at a time systematically documenting every minuscule environmental detail. Although their expeditions often occurred within fifty miles of Southern California's coastal population hubs, sometimes they worked in areas that no humans had set foot in for hundreds of years. Plant enthusiasts call these areas botanical black holes; they probably contain species of plants never documented by Western science. Nicole's team entered such regions on foot with mule trains, or sometimes by helicopter. They contended with bears, rattlesnakes, tarantulas, scorpions, collapsed roads, and clouds of biting insects. One ecologist recounted stumbling upon an illegal marijuana operation, replete with assault rifles slung over the branches of nearby trees. Another described being attacked by a swarm of hornets while traversing a cliff. Conditions were perilous: once, Nicole's team returned to their radios to learn that a wildfire was barreling toward them. To escape, they hefted their heavy gear and ran up a mountain slope in triple-digit heat.

I appreciated the ecologists' devotion to science-based forestry conservation, but I was skeptical—not of the science itself, but of its reception. Science communicators struggled to counter the fossil fuel industry's misinformation campaigns, and no amount of data could convince some Americans that COVID vaccines are safe. Similarly, for those already distrustful of the Forest Service, science might not be persuasive. Facts often don't convince people whose ideological positions are already settled.

In the moment, however, the ecologists' enthusiasm was contagious. We spent our nights cooking veggie stir-fries with jackfruit tacos in lantern light, then sipping IPAs under a star-speckled sky before retiring to our tents. We woke at the first ray of dawn and trudged back to our research plots. Time slowed to a calm heartbeat as the endless routine of counting, measuring, and documenting liberated my mind to attend to the details of the forest. I learned to feel the varied graininess of soil in my fingers, to appraise the height of a tree with my eyes, to pinch a leaf to learn its species, and even to appreciate the tangy smell of ants as they swarmed up my shirt.

After a week of this, I returned home. I hadn't been this dirty or tired since firefighting. The major difference was that I was upbeat now, exhilarated, singing in the shower as I pulled sticks from my hair. We weren't lighting fires yet, but we were working to bring them back. Sustainable forest management might be a distant goal, but I gained a sense of the hard work and dedication that made it even a distant possibility.

Soon after, I checked my email. At the top of my inbox was an invitation from a local conservation group named Los Padres Forest Watch to join the fight against corporate clear-cutting. I squinted at the text in disbelief. The project they were fighting was one in Nicole's jurisdiction.

The project, the Forest Watch claimed, was part of an "alarming trend of commercial logging projects and biomass harvest in the Los Padres National Forest and across the country." According to the email, forest management was a loophole orchestrated by the Trump administration to clear-cut sky islands. All of this, right in my backyard.

I was initially taken aback by the accusation. I knew firsthand that Nicole was a dedicated scientist. Her methodological rigor was paired with a profound sense of personal responsibility for the well-being of the landscapes in her charge. Yet, I realized, the conflict lay in the process. On average, the forests of today hold roughly four times more fuel than they would have held before the era of fire suppression. Biomass translates to energy released by flames. If anyone were to attempt a prescribed burn without first reducing the fuel, the flames would be more likely to get out of control and damage the ecosystems. "When you've got yourself in a hundred-year hole of fuels accumulation," Malcolm North told me, "it's really hard to just drop a match and say it's all going to be good now."

To prevent prescribed burns from becoming extreme, and also to increase the window of opportunity in which burns can be carried out, before fires are lit, the small trees and brush connecting the forest floor to the canopy must often be removed with chainsaws. That vegetation is not "natural" in the sense of "growing without human influence." On the contrary, for over two centuries, our society created this excess vegetation by stamping out any fires that would have kept that vegetation at ecologically balanced levels. These plants have, in a very real sense, colonized the forests as a result of fire suppression. Yet the practice of removing small trees arouses suspicion among those with strong notions of what is natural—chainsaws don't seem to fit the description. "People see right through this," the petition read, "and they're speaking out."

The public piled on, with 99.9 percent of comments submitted to the Forest Service opposing the project as violence against nature. Soon, the removal of ladder fuels was being conflated with wholesale clear-cutting. The mayor of Ojai, a nearby town, condemned the "logging of special and invaluable old growth forests." One commenter pleaded to let "the area burn naturally, as it is evolved to do," while another asked why "natural ecological processes" would "be considered a forest management problem." A Los Angeles resident opposed the project on the grounds that they "feel at peace, at home connected with the flora and fauna, air, dirt, and water." Patagonia, the retail clothing giant, supported a lawsuit to prevent the U.S. Forest Service from cutting down "virgin" forests.

From a historical perspective, the Forest Service had earned the public's distrust. For much of the past century, "land management" was a euphemism for destructive logging practices. True, there was now a new generation of leadership, with Nicole and many others like her doing their part to lead the agency in a sustainable direction. But for anyone who didn't personally know the people planning these projects, there was no reason to abandon the reflexive distrust that had, during many years of true malfeasance, protected cherished landscapes from abuse.

However, the conservationists' claim that "natural" forests should be left untouched was flawed in two ways. First, they tapped into a romantic vision of nature as Eden, reinforcing the loaded notion that forests are virgins, unsullied by society, where humans are intruders and our hands can only render harm. If you follow that seam deep enough, you find it buried in stolen land. The first American settlers of California weaponized the myth of wilderness to justify the seizure of Indigenous land, falsely imagining that land as a place untouched by human history.

Second, California forests are not "natural." From the perspective of Nicole and other scientists, the forests have been influenced by over ten thousand years of Indigenous management, altered by centuries of fire suppression, and are now existentially threatened by climate change. To pretend the forests are natural now, in an era when our carbon emissions have altered everything from the deepest ocean trenches to the highest mountain peaks, would only ensure their speedy demise.

Within the conservation movement, there was one name that I heard more than any other, hissed by almost every scientist and forester I interviewed across California. "*Chad Hanson*," they said. The ecologist who "went rogue." The man who, legend has it, was struck by an epiphany on a hike in the Sierras. He felt there was nothing humans could do to improve nature and proclaimed this with a very loud voice. Hanson was involved with many of the lawsuits designed to stop Nicole and other scientists across the state from saving—or in his view, destroying—the forests of California.

Hanson's colleague, who I'll call Austin, was helping to lead the resistance to the planned prescribed burn project in Nicole's jurisdiction. Austin happened to live in my apartment complex, so I arranged to meet him on a café patio in our neighborhood. I wasn't sure who I was looking for—everyone around me looked like they could be an anthropologist or conservationist—so I presumed Austin couldn't find me either. I texted him that I was with my dog, but another man with a dog sat at the table next to me, foiling my plan. I was so flustered by the anonymous blur of hipsters that I nearly jumped out of my seat when I heard my name.

"You must be Jordan."

"How could you tell?" I asked, struggling to regain my composure.

"You look like you mean business."

This was exactly the opposite of how I was trying to present myself. I was wearing flip-flops. I had ordered my dog a puppuccino. But Austin appeared tense. Emotions run hot around forest management. His hair was light and straight, falling to his shoulders, with a trimmed reddish beard and a crumb in the mustache. His eyes were a startling blue. He drank his coffee out of a reusable Star Wars mug. "So, you're interested in Reyes Peak?" he asked, sitting across from me.

I told Austin I'm interested in conflicts where people are fighting over how to best preserve the forests.

"I wouldn't say we're trying to preserve the forests," he interrupted.

That surprised me.

"That's not how I think about forests," Austin continued. "They're dynamic. They change. It's not about trying to preserve them like a museum."

But if the Forest Service wasn't allowed to do prescribed burns, I noted, many of these forests might disappear before we could get a handle on climate change.

"So we have this misconception that National Forests are just forests," Austin responded. "But in Southern California, it's not like that." He mentioned that 72 percent of Los Padres National Forest is chaparral, which hasn't suffered from fire suppression.

Austin cited a few sources, but he didn't need to. I knew that chaparral thrives with fire every thirty to one hundred years, and if anything, these landscapes have recently suffered from too many fires. The local branch of the Forest Service also knew this, which was why they weren't planning to burn chaparral. They were planning to burn the coniferous montane forests—the sky islands—on the peaks above the chaparral, where centuries of fire suppression have produced wildfire conditions that could threaten their survival.

"Yeah, things get a little more muddled in montane forests," Austin agreed. But then he pivoted, opposing Forest Service projects on the grounds they would be ineffective because they were focused on fuels reduction. "These wildfires are wind-driven, not fuel-driven," he said. "It's extreme weather."

I had come here for an interview, not a debate, but I could not help pointing out that fires are multivariate, which means they are driven by weather, topography, and fuel. Because people cannot control the former two, they focus on the latter. There was "no question," Nathan Stephenson, a forestry scientist working in Sequoia and Kings Canyon national parks since 1979, told me, of the general effectiveness of fuels treatments for moderating extreme wildfires.

That said, Stephenson emphasized that fuels treatments aren't black and white. They are multistep processes that should be adapted to context. Typically, you need to start by stacking surface fuels—logs, sticks, and bushes—into piles, like bonfires in the forest, which you burn in favorable conditions. Then, with the surface fuels reduced, you remove the ladder fuels—the shade-tolerant species that connect the ground to the canopy. With these removed, you can do a broadcast burn, lighting the ground on fire and following the fire as it slithers through the trees.

When I spoke about this process with Malcolm North, the forester at UC Davis, he conceded some ground to the Hanson camp.

Thinning alone, he emphasized repeatedly, is not enough, and should be criticized when it's framed as a stand-alone solution. North was concerned that politicians, particularly on the American right, would seize on the narrative of thinning to justify slashing environmental regulations, which would give them a political win for appearing as if they're managing forests without taking on the risks of actually using fire. In most cases, North said, thinning should be viewed as a tool that makes burning more feasible. It is fire that jump-starts the ecological processes that keep forests healthy, such as soil respiration, carbon sequestration, and nutrient cycling. "You can't thin your way out of this problem," he told me. "It helps. It's a tool. But we need fire."

At that café in Santa Barbara, Austin scoffed out loud at the very idea of fuels reduction. "There's this *obsession* with ladder fuels," he said. "The Forest Service only talks about how to separate the ground from the canopy. But it's not ladder fuels that burn these forests. Ladder fuels are debunked."

Later, I could not find any other scientists, nor anything resembling a scientific agreement in the peer-reviewed literature, that claimed ladder fuels are "debunked." But as Austin was saying this, my mind drifted elsewhere, to a personal experience I had had with ladder fuels on the fireline. I was on a chainsaw, with Smitty as my puller, when a gust of wind pushed a ball of flame toward us. The flames were only three feet high, but they charged like an animal. We stood underneath an old incense cedar. The canopy of the cedar started six feet off the ground, so it would have been safe from the fireball had it not been for the young fir tree—a ladder fuel—growing in its shade. I only had seconds to cut that fir tree before the fireball hit, but I panicked, wedging my saw in the wood. The flames hit the fir and started climbing. I kept trying to cut it as the flames reached for the lowest boughs of the old cedar. I could feel my wrists blistering, my beard singeing. I was too late. Márlon yanked the back of my shirt and pulled me away. We had to run from a wave of heat as the old cedar torched, throwing embers into the canopy of other massive trees, which joined together into a new flame front. Márlon was forced to call a Chinook helicopter for support.

Most prescribed burn specialists and scientists I spoke with had similar anecdotes, which is backed by the preponderance of data. Nathan Stephenson told me that, when the megafires of 2021 reached the parts of the forests he had treated, the flames "fell to the ground." This is similar to how the megafire behaved when it entered the sequoia grove where I had camped with the Los Padres Hotshots, which we burned with care before we left. According to Stephenson, people like to point to treated areas where wildfires nevertheless burn with high severity as evidence that treatments don't work. In reality, he said, it's like wearing a seat belt: fuels treatments increase the overall probability of survival. In Sequoia and Kings Canyon national parks, Stephenson said, in the areas they weren't able to treat "it was a disaster. We nicknamed it Mordor because all the sequoias got killed."

Seated across from Austin, I didn't want to sound like a stereotypical hotshot who thought only tobacco-spitting old-timers with a few burn scars could understand how wildfires behave, but the assertion that ladder fuels don't have an impact seemed particularly naive. I stopped countering his claims because I was gaining an acute sense that science wasn't the issue here. If the Forest Service said fuel reductions could decrease wildfire hazards, their opponents would say these actions made wildfires worse. If the Forest Service said ladder fuels should be removed prior to prescribed burns, their opponents would say they shouldn't. Austin and I were missing something here. I understood the sentiment of environmental writer Jordan Fisher Smith, who, after interviewing Hanson, noted that "talking to him is like entering a parallel world where everything is opposite."

Later, when I called Hanson, he repeated many of these same points, but with more vehemence. The scientists I had interviewed were "liars" and "jokesters," whose narratives were "bullshit." "The whole thing," he said, has "become such a scam!"

One of those scientists, Malcolm North, was Hanson's PhD adviser back in the early 2000s. "I cut off all ties with him," North told me. "I'm not usually like that. I'm usually very supportive of my students." But North drew the line, he said, when he was included as a coauthor on several of Hanson's publications that were called out for looking at "misused data and intentionally putting misinformation into papers." For most scientists, that may happen once or twice in a career, but North said that with Hanson's publications it was continual. Hanson was estranged from the scientific establishment, and on the phone with me, he was fired up.

I had heard all Hanson's arguments before, so I didn't engage until I felt we had reached the heart of the issue. Hanson clarified that he was not against prescribed burns. He was against the steps the Forest Service took to prepare their burns. And he opposed these steps on the grounds that the Forest Service used contract crews—private businesses—to carry out the manual labor of removing ladder fuels. This created a perverse incentive, in his mind, for the Forest Service to devise projects that were more financially lucrative than ecologically beneficial, while also encouraging businesses to exploit loopholes to cut larger trees, thereby increasing profit margins.

The example Austin pointed to in our local conflict was a provision that allowed contract crews to cut any trees deemed hazardous. These gray areas, he said, produced legal zones of subjectivity that could be exploited by the contracted business. The business could, for example, justify logging old-growth ponderosa pines if they exhibited any signs of disease, even if it was as simple as a growth of mistletoe on a branch. The Forest Service, from his perspective, could not be relied upon to implement these projects because private industry couldn't be trusted to carry them out.

The Forest Service officials I spoke with in California, however, told me this was a false choice. Selling wood and increasing biological diversity aren't always in conflict, especially in former logging forests, which were planted after clear-cuts. Frank Aebly, a district ranger in Mendocino National Forest (much of which was planted like a plantation, "pines in line"), told me that he starts with ecological goals, then evaluates whether he could use private industry to reduce the public cost of accomplishing those goals. For example, if he wanted to increase biological diversity by creating a meadow, he may hire a private contract crew to cut and remove selected trees.

Hinda Darner, a Forest Service fuels manager five hundred miles north of Santa Barbara, whose projects had also been hampered by opposition, acknowledged the validity of conservationists' concerns. She also stressed the importance of oversight to prevent instances of abuse. But, she told me, the intent of that oversight makes a big difference. "There are conservation groups who are genuinely interested in finding common ground, who will go into the forest with us and engage in good-faith conversations about how our project can be improved. We want to hear from them. We listen to them and alter our projects based on their concerns. But there are other groups who view this as a zero-sum game. They won't walk through the forest with us. And they won't settle. They always take us to court."

In Santa Barbara, the groups opposing prescribed burns fell into the latter category. According to court documents, they received "multiple invitations to visit the project site with members of the Forest Service, which went unanswered." Chris Stubbs, the supervisor of Los Padres National Forest, speculated that this unwavering opposition was built into their business structure—they collected more donations when they went to court. "Tell me this," Stubbs demanded with exasperation, "do they *really* believe the things they say?"

It dawned on me that this was almost verbatim what Chad Hanson had asked me of the Forest Service. Each side claimed scientific authority. Each claimed the other was corrupted by perverse financial incentives. And both claimed that lives were on the line. This did not mean that each stance was factually equivalent. But it did shift my understanding of the conflict. The entire basis seemed to revolve around a failure of communication based on irreconcilable distrust.

For Hanson, that distrust was deeply rooted. He told me he had hiked the Pacific Crest Trail in the 1980s, before current logging regulations were in place, and witnessed firsthand the devastation inflicted when the Forest Service caved to the demands of the logging industry. He walked through whole landscapes that had been clearcut. He went to law school in Oregon in the 1990s, specializing in environmental litigation in a region and time when conflicts between logging interests and conservationists were so pitched that they occasionally tipped into violence. The conservationists of the time won hard-fought limitations to old-growth clear-cutting, but it seemed that the experience had embedded within Hanson an institutional distrust that never wavered. Hanson and his followers were still fighting a fight from a different time, place, and context. They were transposing their principles of absolute opposition to a small project in Southern California, where the Forest Service had rarely, if ever, relied on logging revenue, where the agents of the Forest Service were also ardent conservationists with PhDs, and where nothing was slated for clear-cutting.

In spite of my frustration, I found myself appreciating the work of the organizations that employed Hanson and Austin. Aside from countering the Forest Service's prescribed burns, they opposed oil extraction on public lands, prevented private landowners from blocking access to trails, and advocated for endangered species. Their criticism of fuels reduction projects had possibly improved some of those projects' environmental footprints by forcing the Forest Service to tighten the parameters of the trees that could be removed. There was no villain here, just people fighting to make sense of a world in which history has cast institutions of all kinds in a cynical light.

Toward the end of my conversation with Austin at the café, I asked him under what conditions he would support the prescribed burn project on Reyes Peak. "Are there any changes they could make to the project design that would get you to sign off on it?"

"You know, that's a very good question," Austin said, blowing air out of his mouth in frustration. He looked like he was thinking deeply, a thousand-yard stare somewhere over my head. "No," he muttered. Then, with more force, "No. I can't imagine a scenario where the Forest Service would do what's right."

In the end, the Forest Service won in court. In the ruling, the judge noted that Los Padres Forest Watch had based their opposition on false claims. The Forest Watch had rallied the public based on claims that the Forest Service would allow old-growth trees to be clear-cut, that the Forest Service had failed to conduct the necessary environmental reviews, that the Forest Service had ignored the interests of the area's tribes, and that the Forest Service hadn't engaged in a collaborative process with the public. All the claims undergirding the basis of opposition to the project, the judge ruled, were demonstrably false. The judge also spoke of the science, noting that the preponderance of evidence was on the side of the Forest Service. "The USFS relied on the best available science regarding the improvement of forest stand resilience," the ruling affirmed.

The next time I saw Nicole Molinari, I expected to find her jubilant, but she just looked tired. "We won," she told me, "but we also lost." The court battle, she said, had simply increased the bad blood between the agency and the community, ensuring that future projects would receive even more resistance. She was resigned to the fact that she would need to fight for every inch of land in her care.

CHAPTER 15

There are two quick ways to tell if a forest was clear-cut in the past. One is to look at the species. Clear-cut forests are filled with one type of tree, whichever was fetching the highest lumber price at the time it was planted; old-growth forests are more diverse. The other is to look at the age of the trees. Clear-cut forests are younger because they were seeded all at once, after corporations removed the old trees. By contrast, the age of trees in old-growth forests can often be measured by centuries, sometimes millennia.

I learned these facts, and many other things about logging, from a woman I'll call Michelle, the lead forester on state land thirty minutes north of Santa Cruz. I had come here to find out what was slowing prescribed burn operations in the north. Prescribed burns are generally more publicly accepted in Northern California, partly because fire suppression efforts were introduced by American settlers a century later than in the south, allowing the culture of controlled burning to persist. And yet, in spite of more public acceptance, the task of expanding burn operations was proving difficult.

Michelle drove me in her truck down a steep dirt road into a river valley. She wanted to show me the two-hundred-year history of this forest. The redwood trees around us had never been cut. Their enormous trunks were the same russet tone of their inland sequoia cousins. I couldn't see their crowns because they disappeared in mist several hundred feet overhead. The redwoods muted the sounds of the world until all I could hear was the swish of branches and the gurgle of a nearby creek. I felt less like I was inside a forest than inside a timeless creature. But before these romantic notions could carry me away, Michelle brought me back to earth.

"This is the only old-growth stand in our forest," Michelle told me. "It's where the logging executives used to bring their families for Sunday picnics. They liked the trees, so they saved some for themselves, but they cut everything else." In 1850, redwood forests like this had covered around two million acres of California's coast. After just seventy years of American logging, they were reduced by around 95 percent. We were in one of the few groves that remained.

Michelle was middle-aged, with strawberry-blond hair and a focused, no-nonsense manner that I eventually came to see as common among foresters. "The logging here changed everything," she told me, tromping down a slope of pine duff until we stood on the bank of a stream. "After they cut the trees, they hauled the logs out down this riverbed. You see how it flows straight down now without any pools or eddies? That's because the logs tore out the rocks and natural dams. The water flows straight out now, so there's less for the trees, and nowhere for fish to spawn. You can't look at anything here without thinking about the logging history."

To prove her point, Michelle wanted to give me a view of the whole forest. We walked back to her truck and drove up the face of the mountain, following a twisting dirt road. At the top, emerging onto a bald peak, I felt like we had come out of a tunnel. The sunlight was temporarily blinding. The sounds of the world washed over me again, airplanes and highway traffic and the yips of mountain bikers. Michelle led me to the edge of a cliff, where I could see the entirety of the forest in her care, a jagged mountainous horizon set against the pale haze of the Pacific Ocean.

"Everything you can see here was clear-cut back in the 1930s," Michelle said. It was a mistake, she told me, to think that the logging industry had always been opposed to fire. They often used slash-and-burn tactics to remove the trees. Loggers would go into the forest to cut down the largest, oldest, most profitable ones, then torch the rest so they could haul out the logs, which were so thick that the flames didn't damage them. After, they would reseed the desiccated slopes with whatever species of tree was most profitable at the time. "That's the point where they wanted fire suppression," she told me. When their plantations regrew, the industry lobbied for full suppression so their assets wouldn't burn.

Up until this point, I was prepared to view all loggers as villains who killed good fire. Their politics of total fire suppression occasionally continued into the present—at least one prescribed burn in the county had allegedly been blocked recently by the aging executive of a logging company who was afraid the fire would escape and damage his crop of redwoods. I was surprised when Michelle told me she was not opposed to logging, that it was necessary for the survival of her forest.

She asked me if I remembered the first stretch of forest she had taken me to. I did. It was a lush, sun-dappled slope filled with young redwoods, ferns, and spruce. That area was her baby, she said, because it was where she had personally been logging. It was special to her because she walked through and, by hand, chose every tree that needed to be cut down. Those decisions allowed her to actively design the character of the forest. She chose which trees to cut based on a scientifically intimate understanding of the ecosystem: the density of the trees, the age of different plants, and the species she wanted to encourage. The difference was that Michelle logged not to make money but to maximize biological diversity and improve overall forest health.

If Michelle's forests had never been clear-cut, perhaps logging would not be necessary. In that imaginary world, she would have been able to maintain the health of the ecosystem with fire alone, burning periodically to keep the habitat in balance. And maybe she could get to that point someday if she were able to reintroduce fire to every habitat where it was needed. But Michelle, like most foresters in California, inherited the remnants of a plantation. For her, this did not mean that the forest was less valuable, but that she needed to actively work to transform it back into a dynamic, healthy place. This meant removing some trees where they were too dense to all thrive, balancing the distribution of species, and reducing the amount of fuel so that fire, when it inevitably arrived, enhanced the health of the ecosystem. And all of this, for Michelle, required logging.

The category of logging, she thought, needed more nuance. Just as some kinds of fire are regenerative and some destructive, some kinds of logging can be beneficial and some damaging. In Michelle's case, it helped that her logging operations weren't for profit, which allowed her to avoid the pressure to keep cutting more. Instead of becoming caught in a cycle of extraction, whatever money she made from logging went toward paying for the operations of her forest: the trails, research, and soon, prescribed fires.

Initially, Michelle encountered many of the barriers that plagued the rest of California. When she arrived a decade ago, conservationists were actively mobilized against all logging in the area, and understandably so. But instead of fighting with the conservationists, Michelle spoke with them, inviting them to the reserve so they could understand how and why she cut down some trees to cultivate others. She took time to build relationships and develop trust. Now, conservationists didn't oppose her projects. "People view forest management as dealing with trees," Michelle said, "but so much of it is dealing with emotions. Those emotions are usually valid, and I take them very seriously."

Back on that ridge, squinting through the sunlight, I was hyperaware that we stood on an island of unburned forest. Nearly every other forest in the region had burned two years before. Someday, this would burn too. Looking at the trees below, I cast about for optimism. Amid the bluish haze, I could discern lighter colors, the emerald green of redwoods that had survived industrial logging and now rose like sentinels over the rest. I found myself saying aloud that maybe, if a forest could grow back from such industrial devastation, it could come back even if a wildfire wiped it all out.

"I'm not so sure," Michelle admitted. "Everything is different now. The baselines of our planet have shifted just in the time since the loggers planted these trees. So we don't know what would grow back. That's a big unknown."

Michelle told me she used to think a lot about climate change. It kept her up at night. But now, she said, her work helped her find peace. "This is my forest," she said. "And I'm doing everything I can to ensure that, when it burns, it survives intact. This is one small piece of the problem, but it's something I can control."

A few months later, Michelle lit the first fire this forest had seen in about a hundred years. It was a small fire, a mere fifteen acres—just over half a percent of the land in her charge—but she had been planning it for a year. She needed the burn to be perfect because this was a pilot fire, designed to make everyone in the county comfortable. Afterward, she would expand. And there were others in the county with their torches on standby, waiting for Michelle's fire to succeed so they could begin lighting their own.

I asked how much she wanted to burn.

She smiled. "I'm going to burn it all."

※

LEAVING MICHELLE'S FOREST, I DROVE TOWARD THE COAST AS THE SUN began to set. The trees whisked by, blurring in the dusk. I parked above the cliffs, found a path down to the beach, and followed it to a slab of rock that rose from the low tide. The rock looked like a fossilized whale, reaching far into the ocean. As I approached it, my feet crunched mussel shells like shards of glass. I felt kelp pop under my shoes. The rock was bleached white from salt, waves crashing, leaping, grasping, receding. A dead seal was pressed against the base of the rock, scapula protruding through rotting parchment skin. Orange plastic was wrapped around one of its fins.

I climbed onto the rock and followed it out over the ocean. The wind blew from the north, cold against my neck, pressing my windbreaker to my back. I reached the end of the rock and stood on its edge, looking into an abyss of darkening ocean and reddening sky. I turned to look at the hills. Where they met the ocean, they appeared cloven as if from a knife, falling away. The cliffs were lined by a forest. The sound coming from that forest was one I hadn't heard since the fireline: a phantasmal, persistent moan of air rushing through bare bough and limb. As far as I could see, the edge of our continent was shrouded with trees that were black and dead from a megafire.

I turned away from this ghost of a forest. The horizon was gone now. The night sky met a rising sea. I hoped Michelle would succeed.

※

IN THE SPRING OF 2022, RESIDENTS OF LAS DISPENSAS, NEW MEXICO, stepped from their homes expecting to see snow in the mountains. Instead, a flickering mushroom cloud towered over them. The Forest

Service had been carrying out a prescribed burn. It escaped. The ensuing megafire would consume over three hundred thousand acres and become the largest in state history.

Not twenty-four hours after I saw this fire in the news, my phone started pinging with videos from the Los Padres Hotshots. They were leading the night operations—night was the only time the fire was calm enough to approach—and they were looping a lasso of prescribed fire around the wild one's head. The megafire was so large that it took the hotshots two weeks to finish, burning along a line that stretched some forty miles. They worked from dusk until dawn, sunset to sunrise, which meant they were expected to sleep on the ground under the sun in the heat of the day. Drogo sent me a video of Aoki napping peacefully with flies crawling on his face. But Drogo grew so tired that he began to hallucinate. At the end of the two weeks, he found himself shirtless in a field under the blazing sun, holding an ax. He had thought he heard a cow yowling, keeping him awake, so he had set out to kill it. He came to his senses in the midst of this stupor, saw the ax in a hand that didn't feel like his own, and ran back to the buggy, where he curled into a ball with a shirt over his head and trembled until night came again. This was Drogo's last season on the crew.

The Los Padres Hotshots contained that fire, but this was not the message that circulated. In the area, many locals believed the megafire was intentional. "It is still mind boggling," a local man wrote on social media, "that the Forest Service might be burning all of our beautiful land on purpose. I've been posting about it." For the broader public, the message was that inept government agents had mistakenly sparked a natural disaster, and that it was criminally reckless to try to burn in climate-changed conditions. Soon after, in Oregon, a local sheriff arrested a Forest Service burn boss after his prescribed fire briefly escaped and burned a fence that demarcated private property.

In a sense, public inhibitions about prescribed burns in a climate-changed era are valid. Climate change is self-fueling. It makes many solutions more difficult to implement even as it makes them more necessary. Prescribed burns can increase forests' resilience to climate change, but the window of time in which people can safely carry out prescribed burns becomes less predictable and shrinks every year. And even isolated mistakes threaten to undermine the larger effort.

Climate change presents less direct obstacles as well. Many regulations and protocols were written for a cooler planet with a stable climate. In Southern California, for example, regulations prevent prescribed burns from being lit in the spring after the oak trees have begun to bud. These oak trees now bloom weeks earlier than they did just a generation ago, eliminating the already narrow window of time in which foresters are allowed to burn.

Even when prescribed burns are possible, the carbon math doesn't intuitively add up. In California, the carbon emitted from the 2020 wildfires erased that year's progress in reducing emissions. In 2023, wildfires in Canada released roughly two billion tons of carbon—triple the annual footprint of the entire country. Fires are even beginning to transform the Amazon rainforest from a carbon sink to a carbon source. Around the world, forests are nearing tipping points where they become net contributors to climate change, emitting more carbon than they consume. Faced with this carbon math, the solutions can seem deceptively simple. Keep carbon out of the air. Or as the title of a Bill McKibben piece suggests, "In a world on fire, stop burning things."

Recent research, however, suggests that burning landscapes the right way can actually keep carbon out of the atmosphere. These results came after a team of scientists analyzed the impact of prescribed fire on soil. Globally, soil holds more carbon than both vegetation and the atmosphere combined. In places where the types of fire match the needs of the biome, the scientists found, carbon storage in the soil can either stabilize or increase. In some instances this occurred because the nutrients in the ashes increased the biomass of root systems; in others, it was because the fire created clumps of charcoal in the soil, which slowed decomposition and carbon leakage. Overall, the soil carbon captured by appropriately burning can outweigh the carbon released as smoke. "Ecosystems can store huge amounts of carbon when the frequency and intensity of fire is just right," said Adam Pellegrini, a professor in the University of Cambridge's Department of Plant Sciences. He hoped the research would reinforce the idea that "when managed properly, fire can also be good—both for maintaining biodiversity and for carbon storage."

Of course, these benefits only come when prescribed burns go as planned. Escaped burns often have the cascading effect of slowing progress everywhere. In New Mexico, the escaped fire triggered a ninety-day ban on any prescribed burns in national forests across the entire country. Lenya Quinn-Davidson, a fire adviser for the University of California Cooperative Extension, thought this ban was wrongheaded. Counterintuitively, Quinn-Davidson believed that the escaped fire highlighted the necessity of increasing the number of fires on the land. "It's not just about how that fire started," she said, "but the condition of the landscape it was burning in."

The Forest Service claims a prescribed burn success rate of 99.84 percent. Of the 0.16 percent that do escape, Quinn-Davidson pointed out, even fewer cause damage. "Why are we spending so much time focused on the things that go wrong when almost one hundred percent of the time it goes right?" she asked. "We know that prescribed fire is our best tool to prevent these catastrophic wildfires, but we keep adding restrictions making it harder to do it."

It may be difficult to overcome the public's fear of prescribed burns, but there are ways to mitigate the financial and legal risks to those conducting them. California, to encourage more prescribed burning, has implemented measures to transfer some liability from individual burners to the California Department of Forestry and Fire Protection, while also creating a state insurance fund so that burn bosses can access up to two million dollars in support for damages if a prescribed fire escapes. For an individual to be liable for suppression costs on California's state and private land requires proof of gross negligence.

Adequately funding the Forest Service would further aid prescribed burning. Most of the land that needs to be burned is federal. To ensure that these burns proceed safely would mean funding not just the projects themselves, but the people in the federal government who are implementing the projects. Due to low wages and poor working conditions, the people with the expertise to safely conduct prescribed burns are fleeing the Forest Service at a time when they are needed most. This lack of funding doesn't just imperil the hotshots, or the ecologists, or the forests under federal jurisdiction, but everyone with a stake in our global climate. When prescribed burns escape, it would be wise to ask why people with such grave responsibility are often being paid near minimum wage.

Most people aren't asking this. The Los Padres Hotshots were able to stop that fire in New Mexico, but Drogo texted me that, in towns where hotshots had once been met with applause, they now received hostile glares. He watched a mother shuffle her children away when the crew entered a restaurant. A local sheriff requested that another crew of hotshots camp outside the city limits, where they wouldn't disturb the population. They were federal agents, after all, doing the job of the public sector in an era where time, money, and trust were all in short supply.

THE IDEA TO ALLOW OUR PRESCRIBED FIRE TO ESCAPE OCCURRED TO A man I'll call Bruno as soon as the air quality officer showed up. It was a year to the day after I had been released from the hotshots, and I had joined an assorted group of people working to return healthy fire to California's ecosystems. There were forty of us in this coastal woodland—scientists, ranchers, conservationists, firefighters, and tribal representatives from across California, Canada, and Australia—who had all converged to spend the week setting Santa Barbara County on fire. Now, the entire operation seemed to be in jeopardy because of the air quality officer's presence.

Bruno, an experienced burner, fumed in the mottled light of the oak forest. The oaks were giant and gnarled, casting puzzle-board shadows over green grass that plunged over a cliff into the Pacific Ocean. Bruno looked like he belonged here. He had a handlebar mustache, waxed leather vest, and flannel shirt with the sleeves rolled up to reveal tattooed forearms. The week before, Bruno had burned four hundred acres up north with several tribes. The previous evening, campfire light glowing on his face, he had delivered a rousing speech about the need to bring fire back. So today, when the air quality officer decreased our burn allowance by two-thirds, Bruno was looking for other options to achieve his goal.

He wasn't *telling* us to let our fires escape, he clarified. But if the fires *did* escape, he wouldn't be mad. He knew the conditions. There was no chance a fire could move quickly enough or burn hot enough to cause any damage. And the more fire that touched the ground, the better. Prescribed burn enthusiasts sometimes call escaped fires "bonus acres."

Bonus acres sound irresponsible at best, criminally reckless at worst. For people intimately familiar with prescribed burning, however, the idea is not as wild as it seems. "It has to pass the stupid test," Bruno told me. Where he lived, the state parks let him burn without even scraping a fireline. All that was required was enough moisture to dampen the understory. He let the fire burn right up to the woods, and the fire would go out. "Some people make this way too complicated. It's like, you walk into the woods, the woods are wet, not gonna burn."

That day, the grass was wet, there were no ladder fuels, and the weather was cool. We really didn't need to worry about the fire getting out of hand. But there were concerns about the air quality. If it wasn't the air quality, it would be something else. Foresters like to complain that it takes five years of planning to do five days of burning. For Bruno, climate change wasn't really making prescribed burning more difficult. It was the layers of bureaucracy that made it difficult to burn when the windows of opportune weather opened. "We get tons of days in coastal California that are perfect for burning," Bruno said. "The real question is, are you flexible enough to act on it?"

I stood by Bruno and fixed an appropriate scowl on my face as the air quality officer lumbered up the dirt road in a hard hat and untucked yellow fire shirt.

I wanted to talk to him. I asked Bruno the man's name.

"Mr. Mommy-Didn't-Love-Me-Growing-Up," Bruno said.

If you spend enough time in California, you'll notice that it's a sort of dinner-table cliché that heavy regulations prevent sustainable forest management. The sentiment is not entirely wrong. Endangered species need to be protected. Watersheds need to be cared for. Smoke, especially, needs to be managed to maintain a happy and healthy public. I imagined it would be a difficult task to convince the public not only to accept more flames in their landscapes, but more smoke in their air.

I tracked down the air quality officer and found him talking to a man who worked for San Diego County's Pala tribe. They were leaning against a tree looking into the area we wanted to burn. I was surprised to find the regulator bobbing his head in enthusiastic agreement with whatever the Pala employee had just said. "It's really incredible," the officer was replying. "Fire's connected to *everything*, man, everything, the plants, the soil, the animals, everything. And science hasn't even figured it all out!"

I revised my questions as I shook the officer's hand. His name was Paul, and he certainly wasn't the fire-hating villain I was looking for. An Italian American with a mop of salt-and-pepper hair, Paul had teamed up with a local fire chief who also had Italian heritage, and they hoped to get as much fire on the ground as possible. They called themselves the "burn mafia."

If Paul was part of the "burn mafia," I asked, why had he reduced the amount we were allowed to burn that day?

Paul didn't answer the question directly, but explained how complicated the task was. The technical elements were the easy part, he said. If there were winds, he needed to make sure they would blow the smoke away from hospitals and schools. If there was an inversion layer—a warm air cap trapping cooler air below and preventing smoke from rising—he needed to make sure we lit the fire hot enough for the smoke to punch through, dispersing in the sky above. This meant he needed to understand the topography, the types of plants, and how moist they were. That was the easy part.

The complicated part of Paul's job as the air quality officer was managing the social drama surrounding prescribed burns. When these projects were shut down, the process often happened through back-door dealings. In Santa Barbara, anyone angry about a burn project had two avenues of recourse. They could complain to the county firefighters, but they were notorious pyromaniacal ruffians who rarely gave an inch. So angry citizens sometimes called Paul. In Paul's world, prescribed burns could unfold like a medieval court drama. "I was trained as an economist," he told me, flapping his arms, "not a diplomat."

I asked who would oppose a prescribed burn.

Paul shrugged, dodging the question again by telling me who *wasn't* opposed. The public was surprisingly amenable. Not a single complaint this year. Likewise with the county government. But I knew there were plenty of other possibilities. California was a crowded place with a lot of concentrated wealth. The villain could be Elon Musk. He was scheduled to launch a rocket into space that evening. The launch pad was a few miles north of us. Musk's employees, Paul said, were monitoring us with a drone. Or it could be the conservation groups. They had delayed some burn projects to the south. Or it could be the neighbors, but I had spoken with them, and they were friendly. I had other suspicions, stoked by rumors muttered late at night, after a few beers, under the burners' breath before we retired to our tents. The villains, they said, were in the vineyards.

In recent years, vineyards have been particularly traumatized by megafires. The conflagrations of 2020 cost the industry $3.7 billion, conservatively. People went out of business not because their grapes were scorched, but because they were exposed to smoke. Smoke is carcinogenic, so when it blows over a vineyard, it triggers a defensive mechanism in the grapes. The grapes contain enzymes that bind sugars to the invading particles in the body of the fruit before those particles can reach and damage the seeds. This traps smoke within the grapes. Grape growers call it smoke taint, and connoisseurs don't like the taste.

Personally, I thought this was nonsense. I love a smoky beer, or the occasional peaty scotch. Why couldn't wine adapt for the greater good? From an anthropological perspective, taste is among the most subjective of the senses. We can convince ourselves that nearly anything is tasty if we associate the taste with status and identity. Before lobster was a luxury, it was prison slop. Turophiles drool for reeking Camembert. People enjoy *cigarettes*. Surely smoky wine could be a hit. It was all in the marketing.

When I floated this to Alisa Jacobson, a local grape grower and founder of the West Coast Smoke Exposure Task Force, she listened patiently. She had recently returned from a vacation in Indonesia, and she listened to my naive speculations with the countenance of a woman who could relax enough to free dive with whales. Then, as kindly as possible, she informed me that smoky wine doesn't taste like a barrel-aged stout. "It tastes like—well, it replicates the molecular compounds of fecal matter," she said.

"Really?"

"Yes."

"You're telling me smoke makes wine taste like shit?"

"Yes. Or an ashtray."

"Well, that's not very subjective."

"No. No, it's not."

Like many grape growers, Alisa was an environmentalist. She understood the benefits of prescribed fire, but she was also concerned for the livelihoods of grape growers. Ideally, most prescribed burns would be completed after the harvest, but currently harvest season was when most controlled burns take place in California.

I asked Alisa if vineyards differentiated between different kinds of smoke. After all, the brief white fog produced by prescribed burns was categorically different from the weeks of dense, black smoke that settle over California during wildfires.

Alisa acknowledged this point. "A lot of the conflict comes from a lack of information," she said, "which creates a fear of unknown effects." The science of smoke taint is in its infancy. Will one hour of exposure damage a crop? Or does it take a week? The industry wants clear, easy answers. They want a quantifiable impact on a scale of one to ten, with smoke exposure tied to the sensory experience of taste and smell. Without settled science, Alisa told me, malicious actors exploit the unknown. Wholesale grape buyers often demand lower prices if prescribed burns have happened in the area, claiming the wine will be of lower quality. In the long run, prescribed burns may reduce wildfire hazards, but in the short term, for many grape growers, they're not worth the risk.

I pushed back on this. In California, fire was inevitable. The question was not if it would come, but when. The goal of prescribed burns was to bring fire to the landscape in conditions that were favorable to both the ecology and people inhabiting the area. Surely even grape growers could acknowledge this.

"Theoretically," Alisa conceded. Grape growers were intimately familiar with their ecosystems. Everything from the soil chemistry to the local flora influenced the character of the vintage. They knew smoke could be managed more effectively on prescribed burns. But to support a prescribed burn would require *trusting* the people doing the burn to manage it correctly. As it stood, Alisa told me, grape growers really didn't have much of a say in where, when, and how prescribed burns happened. "It's a point of tension," she said, "and a recipe for a lot of angry people." She went on. "Cooperation would require developing trust. Trust requires relationships. And relationships take time to build."

These dynamics scale up. The difficulty faced by vineyards also represents every other industry impacted by smoke and fire in California, from hospitals, to sporting events, to tourism. The entire economy of California grew in a fire vacuum, and it is struggling to adapt to this new reality.

These dynamics also trickle all the way down. At that prescribed burn in the mottled light of the coastal oak forest, Paul, the air pollution control officer, told me he was working to build relationships between Santa Barbara's grape growers and the county fire department. But there were more immediate relationships on his mind. Not only might the vintners not trust Paul to keep our smoke under control, but he didn't yet trust *us*. This was, in part, why he had decreased the amount we could burn.

A truck rolled up to us. Through the windshield I saw the beaming face of our burn boss. He was a contractor who traveled all over California to burn, and he was accustomed to people not trusting him. Paul climbed into the passenger seat. An hour later, they came back. Paul was laughing. They appeared to be fast friends.

Paul doubled the size of our burn area. We lit our torches and got to work.

BY THE AFTERNOON, SMOKE WAS DRIFTING THROUGH THE OAK FOREST and rustling the leaves. The grass took on the color of roses. Particles of ash darted around us like fireflies. After my years of fighting fires, I had expected the flames to spark my adrenaline. Instead they coaxed me into calm. I sat beside Harry, an aging First Nations Nlaka'pamux man who had driven down from British Columbia to participate in this burn. Our backs were against an old oak. I had scraped a semicircle of dirt around us so we could sit and eat lunch while we watched the line of flames.

The contrast to my experiences with megafires was striking. There was no danger here. The flames were gentle and calm, fluttering just a few inches off the ground. They would stop at the dirt line that surrounded us. We could easily step over them if we needed to move.

Harry wasn't watching the fire. His eyes were on the sky. He uttered a soft, joyful sound. "Now isn't that something?" He pointed above us. "Red-tailed hawks. They always follow our fires."

"Why do you think?" I asked, watching the broad-winged birds ride the drafts.

"Who knows? Maybe there's insects up there. Maybe they're looking for a meal." He chuckled. "Maybe they just like to play."

The truth was, no one really knew the details of how the forest was responding to our fire. A scientist from Australia was measuring the soil to learn how fire affected the microbial life beneath our feet. A pair of NASA researchers lugged around a machine to document the microbiome of the smoke. A PhD student was tracking how spores in oak leaves responded to the flames. An ornithologist was trying to understand how smoke affected the respiratory systems of birds. A geographer was using a drone to track plant regrowth. Our seat at the edge of this fire was a seat at the edge of the scientific unknown.

I felt comfortable on that edge with Harry. He had been a firefighter for his First Nation community for fifty years. He reminded me of my late grandfather, with a tan face and eyes crinkled with perpetual amusement. He had a way of pulling me into his internal calm, speaking with a stream of consciousness that usually ended in a punchline, even when dealing with serious matters. In this lighthearted fashion, Harry emphasized that California's broken relationship with fire was not ours alone. The colonialism and genocide, corporate seizure of land, criminalization of fire, and now, climate change—these building blocks of megafires could be found across the planet. The histories were shared because they were embedded in the shared trauma that built the structures of our society. But if the history was shared, the solutions could be as well.

"It starts here," he told me. "It starts in our relationships with each other. With the land. These things aren't separate from each other." This was why Harry had driven over a thousand miles from Canada to Santa Barbara. He wanted to find common ground, to share his knowledge, and to learn from us. Maybe, if we could work together to bring fire back to California, he could help bring it back to his land in Canada as well. "Everyone has a sphere of influence," Harry said. "We'll need to build relationships with each other if we want to build solutions."

Harry told me about his own grandfather. How when his grandfather was a child, the Canadian government had taken him from his family and locked him in a Christian boarding school. His grandfather wasn't allowed to speak his language. Harry told me how his grandfather had collaborated with other Indigenous children to keep the language alive in secret and, with it, knowledge of fire and the land. Then that land was taken by logging corporations that banned fire and forced its suppression. Without fire, the important plants also began to disappear.

But Harry remembered a day when he was very young and his grandfather very old—a day before our climate changed, before fire suppression loaded the forests with fuel. On that day, his grandfather took him for a walk. They ambled down a dirt path toward an alpine lake, passing through a meadow awash in flowers and vibrating with insects. His grandfather carried a newspaper under his arm until they neared the lake. Then he struck a match. He lit the paper on fire, stooping slowly. He allowed the flames to lick the grass beside the trail until the grass curled and steamed and accepted the flames. Harry stood with his grandfather, watching the fire roam through the meadow like a gentle creature. The plants passed the flames from one to another in a torchlit procession.

Harry, as a child, couldn't yet read the landscape as his grandfather could; he couldn't identify the arc of the wetland that would naturally put the fire out. "Grandfather! Why'd you do that?" he asked.

His grandfather told him he lit the fire for the land. Helping the land meant helping the deer. And helping the deer meant helping himself and his community. "I'm old," Harry's grandfather quipped. "I can't chase the deer anymore. I need them to come to me. When we light the fire, the deer always come."

Harry smiled as he relived that memory, watching the hawks. "Returning to tradition doesn't mean returning to the past," he told me. "It means reconnecting with our ancestors. It means taking care of the land. When we take care of the land, the land takes care of us. That's what my grandfather always told me."

In the oak forest, the wind picked up, hurrying flames through the grass. The smoke cast crimson shadows. The shadows raced across the land as if carrying some invisible form.

The fire was beginning to worry me. I stood, hefting my shovel, ready to scrape a line around the fire's head to stop it. Harry chuckled again, patting the oak he was seated against.

"You know, Jordan, why don't you sit down and let it burn? Let it burn just a little longer."

CHAPTER 16

IN THE EARLY DAYS OF SPRING I DROVE NORTH, TOWARD CLEAR LAKE. The hills were lush. The wildflowers were so colorful that the vales appeared filled with bright paint. After I passed through San Francisco and crossed the Golden Gate Bridge, I saw fewer Teslas and more pickup trucks. Drivers lifted index fingers in greeting. Addresses were spray-painted on plywood with arrows pointing toward homes. I stopped for a beer in the town where the U.S. Dragoons had ridden toward Clear Lake to initiate California's genocide. In the bar, I told a wealthy local rancher that's where I was going. He said, "They ain't human up there."

Lake County is home to the Pomo tribes. It has been for around ten thousand years. Before Europeans and Americans arrived, the Indigenous people of this region burned an area the size of Connecticut. In 1847, two American ranchers enslaved the Pomo and Wappo people. When they freed themselves, the American government retaliated by funding, supporting, and orchestrating the murder of California's Indigenous people. During this same period, the government established reservations for the Pomo tribes around Clear Lake, next to a mine that extracted mercury and sulfur from beneath the shores. In 1988, the Environmental Protection Agency designated the mine as one of the nation's most toxic sites. Today, Lake County is among the most economically depressed in California. When I visited, the only official mention of the genocide was a small historical marker off the side of the highway. Google Maps led me toward a town named for the slaveholders. In the past decade, catastrophic wildfires have incinerated over 60 percent of the county's landmass. The landscape was more than a burn scar. It looked like a crumpled newspaper slashed with scissors, destroyed as far as the eye could see.

I was there to learn from the Tribal EcoRestoration Alliance (TERA). I had reached out to them after reading that they were hosting an event to train fire professionals in the methods and mentalities of Indigenous fire stewardship. Their effort reflected a broader trend. Across California, Indigenous tribes who for the past several centuries have had their burning practices suppressed are at the forefront of bringing them back. In 2021, after decades of advocacy from Indigenous leaders and activists, the California legislature recognized cultural burning as a legitimate and legal practice. The Karuk and Yurok tribes conduct annual prescribed burns on their lands near the border of Oregon. Federal land managers are also beginning to collaborate with Indigenous people to incorporate cultural fire into their management plans. In Yosemite National Park, the Southern Sierra Miwuk Nation and the Tuolumne Band of Me-Wuk Indians partnered with park personnel to use fire to encourage the growth of native plants. When I attended an otherwise sleepy fire management conference, the panel hosted by Indigenous fire practitioners was packed. Every chair and most of the floor space was occupied. Government agents from the Forest Service and Cal Fire were seated cross-legged on the ground to listen to Indigenous perspectives on fire.

Even in Santa Barbara, the place where colonizers first suppressed fire over 230 years ago, the Barbareño and Santa Ynez bands of Chumash Indians were working to revive their knowledge and practices of fire. "Doesn't look like much, but you've gotta start somewhere," Diego Cordero, the lead technician at the Santa Ynez Chumash Environmental Office, told me in the native plant garden. As he spoke, he was bathing freshly seeded soil with fire from a blowtorch. And he was right. A year later, Chumash elders, for the first time since the colonial period, lit fires in the wetlands behind my apartment. "We're active in cultural revitalization, language revitalization, and doing this burn is one of those missing puzzle pieces," said Robyne Redwater, a Chumash activist.

I was hoping the Tribal EcoRestoration Alliance could help me understand how Indigenous ideas of fire translate into actual practices on the ground. In California, more people are recognizing that Indigenous fire knowledge might help save the state from megafires. However, this growing interest also raises the risk of that knowledge being misused—extracted, misinterpreted, and even leveraged to reinforce the systems that have long oppressed Indigenous communities. Institutionalizing Indigenous knowledge often leads to oversimplification. Authorities trim its complexities to align with narrower state agendas.

In fire management, some worry that the state might reduce Indigenous burning practices to a caricature. This simplification could pave the way for fuel reduction projects that fail to prioritize the specific needs of local people, ecosystems, and landscapes—all of which are central to Indigenous burn practices. Moreover, such projects might not honor Indigenous rights to the self-determined management of their own cultural and natural resources. For many communities, this is particularly problematic. Utilizing their knowledge to prevent megafires and preserve California might actually perpetuate the power of a state founded on Indigenous disenfranchisement—a process at the root of the megafire problem itself. "Saving what for many of our people is a dystopia is not a very good strategy for allyship," says Kyle Powys Whyte, a Potawatomi philosopher and environmental scholar, "because we're trying to get to another point."

I ARRIVED AT CLEAR LAKE AROUND SUNSET. IT IS ONE OF THE OLDEST bodies of fresh water on the continent, pooled against the base of a volcano. I drove for an hour up a mountain, into the burn scar, until I found a place to camp. My sleeping bag still smelled like smoke and sweat from the fireline. I propped my head on a log and drank wine from a bottle and tried to read a book until my eyes closed.

In the morning, snowcapped mountains were pink with sunrise. My hands and feet were numb from cold. The cold had also killed my Prius battery. I walked an hour down the dirt road until I found an old man named James, living in a trailer, who could jump-start my car for me. He had rough hands and a gravelly voice. His home had burned in the most recent megafire.

We pulled up to my car, which James eyed skeptically. "See, this is why you need a truck," he said. "Bigger battery."

"That's what I tell my girlfriend," I lied. "But she likes the Prius."

"Tell her to drive it up the mountain next time."

"Maybe I will."

"It's gonna be a new car or a new girlfriend, tell you that."

I opened the trunk and rummaged around for the battery. I was hyperaware that my back was to James. He stood close behind me. "Don't worry," he said, "I'm not gonna hurt you."

I flinched. If he was sizing me up, he wouldn't be impressed by the fact that, instead of tools in my trunk, I only had dog food, board games, and two sets of binoculars. But it was too late to worry about that.

James dug around in my trunk for a moment, found the battery, and connected his generator. As my car charged, we sat on the hood together, looking into the burned forest. In the space between the dead trees, someone had stacked piles of logs to burn when the conditions were right. Maybe someday whatever life returned to this forest would do so alongside fire. Those burn piles were making me surprisingly hopeful.

"Why'd you come up here anyways?" James asked, interrupting my thoughts.

"Seemed like a place I wouldn't be bothered."

That made him laugh. "You know they found some dead bodies up here last week?"

"You're joking."

"No, sirree."

"How'd they die?"

"Murdered."

"What for?"

"Drugs or women. Seem to be the things that get young men killin'."

"Found them buried?"

"Nope. Bodies were hidden in those burn piles."

※

IN MY REVIVED PRIUS, I ROLLED FROM THE BURN SCAR OF THE MOUN- tains to a different burn scar along the shores of Clear Lake. My encounter with James had reinforced the idea that the damage of megafires runs deeper than the landscape. The well-being of the land is bound up with the well-being of people. The Pomo people of Lake County, like many Indigenous groups in California, say that fire is medicine. When prescribed with the right intensity and frequency, flames can be a healing force. Flames can heal landscapes, but the process of burning can also mend fractures in society and the wounds of emotional worlds. I parked my car near the edge of the lake and walked through an area that had recently been burned, intent on learning how fire heals.

The ground was marshy. The light was blue under low-hanging clouds. The fire that had been lit here had cleared most vegetation, and I was close enough to the water to hear the rattle of reeds, the splash of fish, and a chorus of birds. I arrived in time to join a crowd of forestry officials, policymakers, ecologists, and members of four different California tribes. I found a place between a leader of the Forest Service and a leader of Cal Fire. Together we formed a semicircle around Sara, a Pomo elder, who sat at a table in front of a willow grove, surrounded by baskets she had woven herself. She was there to teach us how to get on the good side of fire.

"In spite of what you've probably heard," Sara said, after introductions, "Native fire management isn't mystical or magical. It's common sense. It means paying attention to your environment. Not all fires burn the same. Whether you're in a pine forest or oak savannah or desert, every biome needs a different kind of fire. It's up to us to pay attention, to learn what fire our biomes need."

Sara used the ground beneath us to prove her point. I hadn't noticed, but we stood on the border of two different burn scars. On one side, the fire had burned too hot, during a season that was too hot. That fire had charred the roots of the tule reeds four inches under the soil, stunting their growth.

On our side of the burn scar, Sara had recently helped younger people burn in a way that benefited the tule reeds. "In this biome," she said, "we want to burn when the soil is moist and heavy to touch, but when the reeds are dry enough to snap in your fingers. That's something you learn from intimacy and experience, not just from numbers and calculations." She made this last point with a chiding tone in the direction of the government officials.

The influence of ecologically appropriate fire, Sara clarified, has ripple effects. On these shores, fires that promote the health of the tule reeds also promote the health of the lake. Tule reeds are connected by rhizomes that spread horizontally underground. Fire promotes new shoots to grow from the rhizomes, which help circulate water and aerate the lake. This process can improve water quality by reducing the stagnant conditions that favor mosquitos and algae blooms. Eventually, the rejuvenated tule reeds also provide homes for native fish and nesting areas for birds.

In the past, tule reeds covered all hundred miles of the lake's shore, but the era of fire suppression, coupled with the impacts of settler colonialism, had almost entirely eliminated them. The lake had suffered along with the species that inhabited it, and the effects of suppression had also spread into the well-being of the tribes. The fibers of tule reeds provided material for their boats, baskets, and dancing regalia, so the threat of the tule reeds' disappearance was also a threat to the continuance of many of the tribes' traditional practices. A man beside me wearing a camouflage hat over a black ponytail identified fresh tule sprouts. They used to spend a whole day driving to find tule reeds, he told me; now they could collect them close to home. If they could get their youth involved in cultural burning, he said, maybe they could also revive their language, which now survived only in recordings.

Sara clarified that the tule reed was just one example of a much broader system. She told us to notice the other plants growing around us, dogbane and dogwood and willow trees. All of these were native plants that had historically been central to her people's livelihoods, and all were emerging from the ground that had burned. Without fire, she said, she wouldn't have been able to weave the baskets on her table, because fire prunes the willow branches and creates the straight shoots she uses in her craft.

To burn in ways that benefited the land required a shift in perspective. "Most of you here think in terms of forest goals. But from the tribal perspective, we don't necessarily think like that. Goals imply an end state. I prefer to think in terms of generation and regeneration. How do our practices impact our plants, our ecosystems, and our own children, one, two, three generations in the future? Remember that we live in holistic systems. Everything we do impacts everything else."

This perspective, Sara explained, shapes how people burn, which in turn shapes the character of the land. Fire is a vehicle that carries our thoughts into the world. The issue, in much of California, is that people are trying to light fires with the same mentalities that they fight fires. During most prescribed burn operations, the burners organize into squads, divide the land into units to attack and defend, and finish by "mopping up"—a term popularized by the Marine Corps to describe the process of hunting down and killing enemy combatants. This military mentality reduces the land to fuel that must be eliminated.

The tribal mindset flips this on its head, said Lindsay Dailey, the cofounder of the Tribal EcoRestoration Alliance. "The first week of training, we focus on building a different perspective. We help people to learn how to see the land through the lens of biodiversity, habitat, carbon, and cultural values. You should be able to look out at a mountain and read the history of fire in that place. Can you read what fire suppression has done there? The arc of history is so clearly written in the land. We're trying to build a lens that allows people to see it."

Once people gain that perspective, Dailey explained, it starts something like a domino effect. You can't learn to read the history of the land without acknowledging what has happened there. During the megafires, Dailey told me, the tribes understood that the destruction was directly tied to colonization, genocide, corporate logging, and the criminalization of Indigenous burn practices. It has been difficult to convince federal officials to adopt this perspective. This creates an issue, because just under half of California's land is managed by the federal government. Any sustainable solutions will require government collaboration with tribes, but these collaborations are difficult to achieve without a shared perspective on the root of the problem. Collaboration, says Frank Kanawha Lake, a Karuk descendant and Forest Service ecologist, "requires humility, truth, and reconciliation. That is a starting point."

In Lake County, the megafires created a new starting point. Even before the rains could wash away the ash that covered the rubble of their homes, the leaders of the local branch of the Forest Service called a community meeting. "They were incredibly humble," Lindsay told me. "They admitted that they messed up. They admitted that they had mismanaged the land. They were honest about their mistakes and asked the tribes, 'Where do we go from here, given that this is the reality we're in?'"

At the gathering on the shores of Clear Lake, I stood next to Frank Aebly, a district ranger for the Forest Service who was charged with managing over a quarter million acres of land. He looked exhausted, with bags under his eyes and wrinkles lining his face. Frank came to work at this forest after getting his PhD in geology at the University of Nebraska. Just two years after he took the post, megafires incinerated 98 percent of the forests in his charge. For Frank, as a newcomer trying to make sense of how to rebuild from the ruins, acknowledging the history that led to the destruction also meant shifting his perspective of authority. His job was to listen to the tribes, he told me, to support and include them. Frank was one of the highest-ranking federal officials in the region, but he stood quietly beside me, listening to Sara speak.

The meeting at Clear Lake made it clear that supporting tribal efforts to reclaim fire was an action, not a passive stance. The year before, California had passed legislation that exempted Indigenous people from many of the rules and regulations that hampered burn efforts in the rest of the state. A woman in the crowd with us, Margo Robbins, contributed to crafting the legislation. Margo was a member of the Yurok tribe, with silver hair and traditional blue tattoos running from the corners of her mouth to the bottom of her chin. She told us that she drafted the policy language in a way that would let all Indigenous people burn, regardless of whether they were federally recognized. During a lull in Sara's speech, Margo announced that Smokey the Bear, the historic, macho symbol of fire suppression under American rule, was retiring. Tribal women, she said, were stepping up to take his place.

Local collaborations between tribes and government officials are at least as important as statewide initiatives. After the megafires, Frank Aebly and Hinda Darner, the region's Forest Service fuels manager—whose great-grandmother burned these same mountains—put the weight of the agency behind the formation of the Tribal EcoRestoration Alliance. The county was among the most poverty-stricken of California, and with so much government money flowing toward forest management and rehabilitation, they wanted to ensure that the money was creating jobs in the community, not being captured by outside contractors. Additionally, in their agreement, Aebly and Darner pledged Forest Service support to burns with cultural objectives, not simply fuels reduction. "I don't think that's happening anywhere else," Lindsay Dailey, the founder of TERA, told me. "One of our biggest successes has been rebuilding that relationship with the Forest Service." This relationship means that they have official approval to burn in ways that align with their values.

It was one thing for government agencies to offer their support to Indigenous burning, but it was another thing for government officials to accept training from Indigenous people. At TERA's first burn, Dailey recalled, the Cal Fire personnel just sat in their fire engines, watching, "not coming out, not introducing themselves." On the second burn, the chief introduced himself, but wasn't super friendly. Eventually, he admitted that the members of TERA knew more about fire than most of his staff, and he invited his whole engine crew to attend the burn.

Megafires emerge from a series of fractured relationships—between fire, the land, our institutions, and one another. The crowd at Clear Lake expressed optimism that these relationships were healing. Theirs was a vision of fire that didn't simply eliminate fuel, but encouraged growth. It was a vision of burning as a political act of resistance to the forces that would reduce the land, and the people in it, to expendable commodities.

In the face of climate change, these relationships are invaluable. Dailey told me that she didn't have hope that burning would solve the climate crisis, or that it would prevent mass extinction and widespread suffering. Instead, she believed it would help people build deeper connections with one another and the land. "This work gives me hope that we can center our community around an ethic of care for the land—that we can work together to heal our relationships with each other and the places we call home."

Seated before us on the shores of Clear Lake, Sara concluded with a story. During the megafire, she had fled. She had been unable to reach her family for nine days. Staring out the window of a friend's apartment, she had gazed toward a wall of flame descending on her home. Her friend, concerned for her, had asked her how she was feeling. "I looked at my friend, and I told her I am feeling hopeful." She hadn't been watching the flames, but the smoke, and she had seen the smoke turn white when it reached the grasses of an oak savannah she knew well. The smoke, she hoped, would purge any parasites that had embedded themselves in the oak trees during the era of suppression. Finally, she would be able to harvest acorns in that forest, just as her own parents and grandparents had done not so long ago.

"I am grateful," Sara told us. "I am grateful that this great destructive force of culture, which has been moving in the wrong direction for so long, is beginning to change its course. We're finally moving in a promising direction."

AFTERWORD

IN THE FEW YEARS I SPENT WRITING THIS BOOK, CALIFORNIA'S CONflagrations have grown more erratic. There were no megafires in either 2022 or 2023. Both years, the total land burned in the state amounted to only around 350,000 acres. By contrast, while I was a hotshot, a single fire burned almost triple that amount of land. In the fire seasons that followed, California's hotshots mostly needed to travel out of the state to find big ones to fight.

There are different explanations for this variance. Most have to do with the weather. In the winter of 2023, California received historic levels of rain. People were kayaking in the streets of Santa Barbara, where more than one-third of its average annual precipitation fell in just forty-eight hours. Beneath the sequoias, in the Central Valley, Tulare Lake reemerged for the first time in forty years. Yet when I asked the hotshots about this, they told me that a wet winter doesn't necessarily mean a quiet fire season. Large fires often begin in fine fuels like grasses and shrubs, which are more abundant following rain. Around

Los Angeles, these anomalous rains were followed by some of the hottest and driest months on record, which turned the vegetation into kindling and fueled a firestorm in January 2025. Thousands of people lost homes, hundreds of thousands were displaced, and dozens died, long after the fire season should have ended. Year by year patterns are becoming even more difficult to predict.

The broader trend is clearer. Even when megafires aren't actively burning, temperatures continue to rise. We're in the middle of a major shift, requiring new ways of thinking about fire and its impact on the environment. One scientist I spoke with likened megafires to a biological system update, syncing flora with the shifting atmosphere. With the accelerating pace of climate change, the biotic world often can't keep up; saplings mature into a climate unlike the one that seeded them. In just over a decade, climate change—the combination of drought, beetle kill, and wildfires—has wiped out over 30 percent of California's Sierra Nevada forests. From a certain perspective, these fires could be seen as aligning ecosystems with an emerging planet— transitioning from forest to brush, brush to grassland, grassland to desert, until there is no life left to burn. Quiet wildfire seasons carry their own subtle danger: the misconception that the problem is solved.

The fires are going to come, but our political choices will shape how they burn. We need to ratchet down the burning of fossil fuels, ratchet up the intentional burning of our landscapes, and support people like the hotshots who work to contain the unfolding disasters of our society's creation. If megafires can remind us of anything, it is of the precarity of our relationship with our environments and the work required to care for those places that matter to us.

Yet despite the need for more resources to address this rapidly changing reality, many hotshot crews are in danger of becoming obsolete. In 2022, one of the top crews in Sequoia National Forest lost so many veterans that they were stripped of their hotshot title and, with it, the elite tasks they were allowed to perform to protect the trees. This situation is prevalent. Across the United States, more hotshots than ever are leaving the career due to poor conditions, inadequate pay, lack of health care, and psychological neglect. The frontline workers best equipped to contain the violent manifestations of climate change are leaving the job.

This process is not inevitable. While it's easy to blame the attrition of the federal wildland firefighting workforce solely on rightwing efforts to dismantle America's public sector, it also reflects a broader political tendency to resist investing in communities most capable of mitigating climate impacts. Policymakers and the public alike often fixate on technological solutions to environmental crises. For example, Silicon Valley billionaires have proposed generating artificial rainstorms to combat fires, and a California gubernatorial candidate suggested using water piped from the Mississippi River to spray megafires. Such futuristic solutions frequently obscure the value of low-tech, human solutions available now. In reality, fire management relies on individuals who have spent decades developing skills born of experience, necessity, and exposure. Those who work with fire are among our society's most finely tuned technological assets as we adapt to a planet that burns like never before.

Before the next fire season, while I was writing this book, I began spending more time in a Forest Service archive; I didn't need the information so much as I craved the company of the hotshots. The archive was right next to their station. In the afternoons, after my research, I began visiting them as an anthropologist and outsider, but also as a friend.

I found the hotshots sharpening their chainsaws. There was a new guy in the barracks. We did pull-ups together. I spoke with Red about his truck and Márlon about his baby. Scheer was gone. He told me he was taking a "year of love" to focus on surfing and photography.

Drogo was gone too. He was using the GI Bill to study finance at an Ivy League university.

Aoki wasn't in the station. As I was heading to my car to depart, he rounded the corner of the barracks, striding toward me, tall and thin, his long hair now more gray than black.

If he was surprised to see me, he didn't let it show. I fell in beside him, almost jogging to keep up with his pace. He was back from his daily hike—just a walk—up the mountain. It looked like he had hardly broken a sweat.

We stopped outside his office. Aoki crouched on the lawn. I joined him. I asked him questions, like I always did. He answered with the curt immediacy of a man who doesn't like abstractions. We carried on this way. Then I asked if he's ready for the next fire season.

This made Aoki pause. He raised an eyebrow. "Well, Thomas, that's always the question, isn't it?"

ACKNOWLEDGMENTS

Writing this book felt a bit like burning alive and being reborn from the ashes. My words can only begin to express the depth of gratitude I feel for all the people who supported me throughout these years of work.

This project would not have been possible without my mentors, friends, and colleagues from the community of fire practicioners. These include Matthew Aoki, Eric Verdries, Hector Medrano, Axel Delgado, Mike Nelson, Alexander Rodriguez, Márlon Gutierrez, Kyle Scheer, Noah Perry, Kyle Von Tillow, Jesse Beamon, Josh Murphy, Robert Lee, Tyler Tomlinson, Jacob Garside, Jacob Alvarez, Sebastian Barba, Johnny Martinez, Paul Hiott, Colin Kennedy, Agustin Gaspartus, and Zaúl Diaz. Thanks also to Lucas Mayfield, Miller Bailey, Diego Cordero, Hinda Darner, Lindsay Dailey, Thea Maria Carlson, Nicole Molinari, and Ryan Fass. Many of these people did not appear in the book due to narrative constraints, but all were fundamental in shaping my understanding of fire.

I would not have had the opportunity to turn my firefighting experiences into a book if it were not for my agent, Alice Whitwham from the Cheney Agency, whose guidance through the literary world has been invaluable. Our artistic visions aligned from the beginning, and our collaboration has been a joy. At Riverhead Books, I am deeply grateful to Courtney Young, my editor. Your advice, from the grand vision to the smallest details, was both subtle and powerful, allowing me to maintain my voice while providing crucial guidance. Thanks for taking a chance on a new author. The quality of the final manuscript was significantly enhanced by my fact checker, William Sydney. I'm also grateful to J. C. Franco, the cartographer, whose map enriches the front pages of this book. Sari Blum, thanks for the photo, and for being such a fun yoga buddy. Additionally, I extend a heartfelt thanks to the entire team at *The Drift* magazine, especially my longtime friend Rebecca Panovka. This book would not exist without you.

Writing can be a lonesome experience, but the process of discussing and sharing my manuscripts made me realize how fortunate I am to be surrounded by a diverse community of artists, poets, and scientists. Julia Fine, you went above and beyond, chiseling commas and digging up dangling participles to push the manuscript past a critical threshold at a critical time. Zoe Sims, everyone deserves a friend like you who can run up mountains while dissecting ideas. Emaline Laney, I cherished our rambles about climate change, life, and everything in between. Thanks also to Elizabeth Floyd, Marshall Sharpe, Noah Dentzel, Miguel d'Arcangues Boland, and Challis Popkey. It takes a lot of trust to send imperfect drafts to friends. This book was made better by the collective contributions of your kind, caring, reflexive minds.

I hold special gratitude for my intellectual community, which has refined my analytical skills and hopefully made me a better person in the process. Specifically, my mentors Michael Wesch, Harald Prins, Francoise Barbira-Freedman, and Barbara Herr Harthorn have guided this journey. My research into prescribed burning was made possible by my collaboration with Nick Williams, Richard Cobb, and Sierra Lyman. I benefited greatly from my conversations with Abi Croker, the rest of the folks at the Leverhulme Wildfires Centre, Josh Lappen, and Liz Carlisle.

Throughout the duration of my research, I was fortunate to be hosted by friends and family nearly everywhere I traveled. Anu Ramachandran and Jack Leupin took me in after I had been sleeping in my broken car in Lake County. Matt and Jess Munz provided a pullout couch and good company during some prescribed burns. Leah Matchett housed me on my drive up north. My soon-to-be sister-in-law, Devin Lujan, and brother, Kevin, gave me a home in the Rockies at the beginning and end of this project. Thanks to the whole Boland family, especially Giles and Judith, for opening your home to me and my writing. You are an inspiration in hospitality.

From the Boland family, I owe a special thanks to Sam. It's a rare quality of friend who can walk all day and talk all night. This book would not be the same without your company, conversations, and seemingly limitless supply of Scotch. Vasili Markou, thanks for all the intrigues. It was a joy to finish my final edits while sailing with you in the Sporades.

Thank you to Kathy Supple, my high school English teacher, to whom this book is dedicated. I would not still be writing today if it were not for your encouragement during those formative years.

I would like to thank my family. My grandfather Don Biggs has provided a lifelong example of what it means to respect science, cherish the environment, and fight for the well-being of others. Thanks to my parents, who have offered unconditional love and support through the years. My brother, Kevin, rode a bicycle down Baja with me while I was training for the hotshots, for which I will always be grateful. Thanks for making me stop for tacos and tecates; next time, we'll take motorcycles. Rachel, you're my inspiration. Lauren, thanks for being my hotshot date and best friend while I was on the crew. Alyssa, thanks for always being there when I need you. I love you all.

Cindy and Alan Wade, thank you for frequently watching the dog while I was away.

I want to give a particularly special acknowledgment to my dog (who requested, in her own way, to remain anonymous) for holding me together during the times I was home between wildfires.

And to Kenzie, of course—thank you for the years you were my home.

A NOTE ON SOURCES

THIS BOOK IS THE CULMINATION OF EIGHT YEARS OF RESEARCH. When I began fighting wildfires, I didn't know I would write a book about it. I did, however, know that I was witnessing and participating in something profound, so I took notes obsessively. While on the fireline, I recorded small details and interesting conversations by scribbling in a notebook. When my notebooks became too saturated with sweat, I would type notes into my phone during breaks. When my phone overheated and wouldn't function, I used memory techniques to store information in my brain for several hours, then recorded it as soon as possible. Each evening, I would lie in my sleeping bag and compile my notes. By the end of my tenure as a wildland firefighter, I had approximately one thousand pages. While writing this book, I cross-referenced my own notes with those of several other hotshots who were generous enough to share them with me. The crew also shared dozens of hours of videos and nearly a thousand photos, which I used in conjunction with official incident reports to triangulate my own memories and recordings. All of this information was supplemented with approximately one hundred hours of interviews with fire scientists, experts, and practitioners. I completed some of the historical work myself with access to a variety of archives. For the rest, I am indebted to the research of many intellectuals and journalists who have spent their careers making sense of the complex intersections between people, fire, and the land. Several character names have been changed by request, and several others at my discretion. Everything in this book is, to the best of my knowledge, strictly factual. If you feel that you have encountered an error, please do not hesitate to report it at whenitallburns@gmail.com. I will review comments and implement corrections in future editions so the book can continue to improve.

NOTES

CHAPTER 1

11 **Fire is unique to our planet**: Andrew C. Scott et al., *Fire on Earth: An Introduction* (Chichester, West Sussex, UK: Wiley Blackwell, 2014).
12 **every society creates a unique fire regime**: Cynthia Fowler and James R. Welch, *Fire Otherwise: Ethnobiology of Burning for a Changing World* (Salt Lake City: University of Utah Press, 2018).
13 **People in southern Mexico**: See, for example, Anabel Ford and Ronald Nigh, *The Maya Forest Garden: Eight Millennia of Sustainable Cultivation of the Tropical Woodlands*, New Frontiers in Historical Ecology, vol. 6 (Walnut Creek, CA: Left Coast Press, Inc., 2015).
14 **Nearly every terrestrial area**: Stephen J. Pyne, *World Fire: The Culture of Fire on Earth* (New York: Holt, 1995).
14 **"The discovery of fire"**: Charles Darwin, *The Descent of Man, and Selection in Relation to Sex* (London: Penguin, 2004).
15 **In Namibia, the Bergdama people**: James George Frazer, *Myths of the Origin of Fire* (New York: Hacker Art Books, 1974).
15 **founding the academic disciplines**: See, for example, Sigmund Freud, *Civilization and Its Discontents* (Mineola, NY: Dover Publications, Inc., 1994).
15 **in the form of cartoons**: See, for example, Kirk DeMicco and Chris Sanders, dirs., *The Croods* (Glendale, CA: DreamWorks Animation, 2013).

15 **humanity's relationship with fire**: See, for example, Michael Chazan, "Toward a Long Prehistory of Fire," *Current Anthropology* 58, S16 (2017): S351–59, https://doi.org/10.1086/691988; also, Nicole M. Herzog, Jill D. Pruetz, and Kristen Hawkes, "Investigating Foundations for Hominin Fire Exploitation: Savanna-Dwelling Chimpanzees (Pan Troglodytes Verus) in Fire-Altered Landscapes," *Journal of Human Evolution* 167 (2022): 103193, https://doi.org/10.1016/j.jhevol.2022.103193.

16 **primatologist Jill Pruetz**: The information in this section is based on an interview I conducted with Pruetz and a review of her published articles. See Jill D. Pruetz and Thomas C. LaDuke, "Brief Communication: Reaction to Fire by Savanna Chimpanzees (Pan Troglodytes Verus) at Fongoli, Senegal: Conceptualization of 'Fire Behavior' and the Case for a Chimpanzee Model," *American Journal of Physical Anthropology* 141, no. 4 (2010): 646–50, https://doi.org/10.1002/ajpa.21245.

16 **early humans would have adapted to**: Jill D. Pruetz and Nicole M. Herzog, "Savanna Chimpanzees at Fongoli, Senegal, Navigate a Fire Landscape," *Current Anthropology* 58, S16 (2017): S337–50, https://doi.org/10.1086/692112.

16 **Many animals that evolved alongside humans**: Herzog, Pruetz, and Hawkes, "Investigating Foundations for Hominin Fire Exploitation."

17 **Pruetz and her colleagues believe**: See, for example, Christopher H. Parker et al., "The Pyrophilic Primate Hypothesis," *Evolutionary Anthropology: Issues, News, and Reviews* 25, no. 2 (2016): 54–63, https://doi.org/10.1002/evan.21475.

17 **cognitive capacity to navigate wildfires**: Fulco Scherjon, Corrie Bakels, Katharine MacDonald, and Wil Roebroeks, "Burning the Land: An Ethnographic Study of Off-Site Fire Use by Current and Historically Documented Foragers and Implications for the Interpretation of Past Fire Practices in the Landscape," *Current Anthropology* 56, no. 3 (2015): 299–326, https://doi.org/10.1086/681561.

18 **safely metabolize carcinogenic toxins**: See, for example, Troy D. Hubbard et al., "Divergent Ah Receptor Ligand Selectivity during Hominin Evolution," *Molecular Biology and Evolution* 33, 10 (2016): 2648–58, https://doi.org/10.1093/molbev/msw143.

18 **By sixty thousand years ago**: Stephen J. Pyne, *The Pyrocene: How We Created an Age of Fire, and What Happens Next* (Berkeley: University of California Press, 2021).

CHAPTER 2

37 **conflagrations that puzzled Omer Stewart**: Scientific investigations into the influence of fire, especially Indigenous burning, on precolonial landscapes are expanding across a variety of disciplines. This book shaped my thinking on the topic and is the source of much of the information in chapter 2. Omer Call

Stewart, *Forgotten Fires: Native Americans and the Transient Wilderness* (Norman: University of Oklahoma Press, 2002).

38 **writes historian Stephen Pyne:** Stephen J. Pyne, *Fire in America: A Cultural History of Wildland and Rural Fire* (Seattle: University of Washington Press, 1997).

39 **wrote historian J. Donald Hughes:** J. Donald Hughes, *American Indian Ecology* (El Paso: Texas Western Press, 1983).

39 **intruders on this state of nature:** John Muir, *The Mountains of California* (New York: Century Co., 1894).

39 **supported their removal:** Mark David Spence, *Dispossessing the Wilderness: Indian Removal and the Making of the National Parks* (New York: Oxford University Press, 1999).

41 **Salish tribes have a Sx'paám:** Adriana Petryna, *Horizon Work: At the Edges of Knowledge in an Age of Runaway Climate Change* (Princeton, NJ: Princeton University Press, 2022).

43 **said Karuk ecologist Frank Kanawha Lake:** Frank Kanawha Lake, "Indigenous Traditional Ecological Knowledge Can Save Our Ecosystems," Bioneers, September 24, 2020, https://bioneers.org/frank-lake-indigenous-traditional-ecological-knowledge-zmaz2008/.

43 **According to Jan Timbrook:** Jan Timbrook, John R. Johnson, and David D. Earle, "Vegetation Burning by the Chumash," *Journal of California and Great Basin Anthropology* 4, no. 2 (1982), https://escholarship.org/uc/item/1rv936jq.

43 **This tightened the snowpack:** M. Kat Anderson, "The Use of Fire by Native Americans in California," in *Fire in California's Ecosystems* (Berkeley: University of California Press, 2006), https://doi.org/10.1525/california/9780520246058.003.0017.

44 **writes Tarahumara scholar Enrique Salmon:** E. Salmon, "Kincentric Ecology: Indigenous Perceptions of the Human-Nature Relationship," *Ecological Applications* 10, no. 5 (2000): 1327–32, https://doi.org/10.1890/1051-0761(2000)010[1327:KEIPOT]2.0.CO;2.

44 **When the land includes everything:** Robin Wall Kimmerer, *Braiding Sweetgrass: Indigenous Wisdom, Scientific Knowledge and the Teachings of Plants* (Minneapolis, MN: Milkweed Editions, 2013).

44 **fire has long been that gift:** Robin Wall Kimmerer and Frank Kanawha Lake, "The Role of Indigenous Burning in Land Management," *Journal of Forestry* 99, no. 11 (2001): 36.

45 **ten million acres per year:** Scott L. Stephens, Robert E. Martin, and Nicholas E. Clinton, "Prehistoric Fire Area and Emissions from California's Forests, Woodlands, Shrublands, and Grasslands," *Forest Ecology and Management* 251, 3 (2007): 205–16, https://doi.org/10.1016/j.foreco.2007.06.005.

45 **The light emanated from flowers:** Kat Anderson, *Tending the Wild: Native*

American Knowledge and the Management of California's Natural Resources (Berkeley: University of California Press, 2005).

CHAPTER 3

55 **aggressive, competitive, and sometimes violent:** John Archer, "Testosterone and Human Aggression: An Evaluation of the Challenge Hypothesis," *Neuroscience and Biobehavioral Reviews* 30, no. 3 (2006): 319–45, https://doi.org/10.1016/j.neubiorev.2004.12.007.

55 **testosterone levels are also:** Sari M. van Anders, "Beyond Masculinity: Testosterone, Gender/Sex, and Human Social Behavior in a Comparative Context," *Frontiers in Neuroendocrinology* 34, no. 3 (2013): 198–210, https://doi.org/10.1016/j.yfrne.2013.07.001.

55 **cause testosterone levels to spike:** Allan Mazur, "A Biosocial Model of Status in Face-to-Face Primate Groups," *Social Forces* 64, no. 2 (1985): 377–402, https://doi.org/10.1093/sf/64.2.377.

56 **Jung, a prominent Swiss psychologist:** See, for example, Carl G. Jung, *Symbols of Transformation: An Analysis of the Prelude to a Case of Schizophrenia, Collected Works of C. G. Jung*, vol. 5 (New York: Harper, 1962).

56 **"a phallic sense":** Sigmund Freud, *Civilization and Its Discontents* (Mineola, NY: Dover Publications, Inc., 1994).

56 **founded the Los Padres Hotshots:** See National Interagency Fire Center, "Hotshot Crew History in America," 2013, https://gacc.nifc.gov/swcc/dc/nmsdc/documents/Crews/NMSDC_Hotshot_Crew_History_2013.pdf.

56 **"the red demon":** The full passage reads: "Today marks the beginning of Fire Prevention Week, a nationwide attempt to curb the havoc wrought by the red demon which last year consumed 393,000 dwellings and took the lives of more than 10,000 persons, a toll for which carelessness was chiefly to blame."

56 **calling for a "firefighting army":** The full passage reads: "Fire is a menace at all times, can destroy assets vitally important to California's welfare when conditions are entirely normal. . . . A peacetime firefighting army as well as one merely for the defense period would seem to be what is needed. The volunteer organization should, therefore, be on a permanent basis."

56 **wrote a 1941 opinion piece:** This quote is from a newspaper clipping I found in an archive in Los Padres National Forest. The author's name was not included in the clipping.

57 **federal edict in 1935:** See E. M. Loveridge, "The Fire Suppression Policy of the U.S. Forest Service," *Journal of Forestry* 42 (1944): 549–54.

57 **first female hotshots:** Her name is Deanne Shulman. She also went on to become the first female smokejumper in the United States. An interview with

Deanne can be found at "Deanne Shulman Interview, July 22, 1984," Smokejumpers 1984 Reunion Oral History Collection, OH 133, Archives and Special Collections, Mansfield Library, University of Montana-Missoula, July, https://scholarworks.umt.edu/smokejumpers/20.

CHAPTER 4

71 **oldest human remains:** See Phil C. Orr, "Arlington Springs Man," *Science* 135, no. 3499 (1962): 219, https://doi.org/10.1126/science.135.3499.219. Also see Todd J. Braje, Jon M. Erlandson, and Torben C. Rick, *Islands Through Time: A Human and Ecological History of California's Northern Channel Islands* (London: Rowman & Littlefield, 2021).

71 **evidence of flames:** See, for example, Mark Hardiman et al., "Fire History on the California Channel Islands Spanning Human Arrival in the Americas," *Philosophical Transactions of the Royal Society of London*, B 371, 1696 (2016): 20150167, https://doi.org/10.1098/rstb.2015.0167. Also see Anna Klimaszewski-Patterson, Theodore Dingemans, Christopher T. Morgan, and Scott A. Mensing, "Human Influence on Late Holocene Fire History in a Mixed-Conifer Forest, Sierra National Forest, California," *Fire Ecology* 20, no 1 (2024): 1–16, https://doi.org/10.1186/s42408-023-00245-9.

71 **Chumash towns dotted the foothills:** Terry L. Jones, "Goleta Slough Prehistory: Insights Gained from a Vanishing Archaeological Record," review of Michael A. Glassow, *Contributions in Anthropology*, no. 4 (Santa Barbara: Santa Barbara Museum of Natural History, 2020), *California Archaeology* 13, no. 1 (2020): 119–21, https://doi.org/10.1080/1947461X.2021.1933002.

71 **While the people inhabiting inland towns:** Lynn H. Gamble, *The Chumash World at European Contact: Power, Trade, and Feasting among Complex Hunter-Gatherers* (Berkeley: University of California Press, 2008).

72 **Soaproot, an innocuous native plant:** Janice Timbrook, *Chumash Ethnobotany: Plant Knowledge among the Chumash People of Southern California*, Santa Barbara Museum of Natural History Monographs, no. 5 (Santa Barbara: Santa Barbara Museum of Natural History, 2007).

72 **When pounded and sprinkled:** Robert O. Gibson, *The Chumash (Indians of North America)* (New York: Chelsea House Publishers, 1991).

72 **The most festive annual occasion:** Hadley Meares, "A Maritime People: The Chumash Tribes of Santa Barbara Channel," PBS SoCal, July 16, 2015, https://www.pbssocal.org/shows/california-coastal-trail/a-maritime-people-the-chumash-tribes-of-santa-barbara-channel.

72 **welcomed the new year:** Dennis F. Kelley, *Tradition, Performance, and Religion in Native America: Ancestral Ways, Modern Selves* (New York: Routledge, 2015).

73 **a ferocious self-torture:** I drew much of this chapter's general historical context surrounding California's colonial period from this book: Kevin Starr, *California: A History* (New York: Modern Library, 2005).
74 **commanded by Francisco de Gali:** José Antonio Gurpegui, "The Coast of California as the Long Projected Hub for the Spanish Empire in the Pacific, 1523–1815," *International Journal of Maritime History* 31, no. 2(2019): 233–47, https://doi.org/10.1177/0843871419842051.
74 **wrecked near Point Reyes:** Matthew A. Russell, "Precolonial Encounters at Tamál-Húye: An Event-Oriented Archaeology in Sixteenth-Century Northern California," in *The Archaeology of Capitalism in Colonial Contexts: Postcolonial Historical Archaeologies*, Sarah K. Croucher and Lindsay Weiss, eds. (New York: Springer, 2011), 39–63, https://doi.org/10.1007/978-1-4614-0192-6_2.
75 **the expulsion of all Jesuits:** Virginia Guedea, "The Old Colonialism Ends, the New Colonialism Begins," in *The Oxford History of Mexico*, Michael C. Meyer and William H. Beezley, eds. (New York: Oxford University Press, 2000).
75 **"you shall suffer the death penalty":** Mariano Cuevas and Peter P. Forrestal, "Expulsion of the Jesuits from Mexico," *Records of the American Catholic Historical Society of Philadelphia* 43, no. 2 (1932): 142–84.
76 **Serra wrote in a letter:** Maynard J. Geiger, *The Life and Times of Fray Junípero Serra, O.F.M.; or, The Man Who Never Turned Back, 1713–1784, a Biography*, Publications of the Academy of American Franciscan History, Monograph Series, v. 5-6 (Washington, D.C.: Academy of American Franciscan History, 1959).
76 **"rats, coyotes, vipers, crows":** Emily M. Smith, Jennifer A. Lucido, and Scott E. Lydon, "Flora, Fauna, and Food: Changing Dietary Patterns at the Spanish Royal Presidio of Monterey, 1770–1848," Boletin, California Missions Foundation, UCLA, 2017, https://escholarship.org/uc/item/078652xg.
77 **"Two or three whippings":** Benjamin Madley, *An American Genocide: The United States and the California Indian Catastrophe, 1846–1873* (New Haven, CT: Yale University Press, 2016).
77 **"The manner in which the Indians are treated":** Lisbeth Haas, *Saints and Citizens: Indigenous Histories of Colonial Missions and Mexican California* (Berkeley: University of California Press, 2014).
78 **"burnt in some spots":** Much of the Spanish correspondence regarding Indigenous burning in this chapter comes from this source: Jan Timbrook, John R. Johnson, and David D. Earle, "Vegetation Burning by the Chumash," *Journal of California and Great Basin Anthropology* 4, no. 2 (1982): 163–86.
79 **One man suffered five whippings:** Benjamin Madley, "California's First Mass Incarceration System: Franciscan Missions, California Indians, and Penal Servitude, 1769–1836," *Pacific Historical Review* 88, no. 1 (2019): 14–47, https://doi.org/10.1525/phr.2019.88.1.14.

79 **Many were tortured and executed:** For examples of the methods used, see the journal of Vassili Tarakanoff, a Russian who was held captive by the Spanish. While some details are difficult to confirm, they corroborate many other accounts from that period. Vassili Petrovitch Tarakanoff, *Statement of My Captivity among the Californians*, Early California Travels Series 16 (Los Angeles: Glen Dawson, 1953).

79 **Sexual abuse of women and children:** Madley, *An American Genocide*.

80 **The mortality rate for children:** Robert H. Jackson, "The Dynamic of Indian Demographic Collapse in the San Francisco Bay Missions, Alta California, 1776–1840," *American Indian Quarterly* 16, no. 2 (1992): 141–56, https://doi.org/10.2307/1185426.

80 **the ground has been watered [with blood]:** Elias Castillo, *A Cross of Thorns: The Enslavement of California's Indians by the Spanish Missions* (Fresno, CA: Craven Street Books, 2015).

81 **"one of the founding fathers":** Gerard O'Connell, "Pope Hails Junipero Serra as 'One of the Founding Fathers of the United States,'" *America*, May 2, 2015, https://www.americamagazine.org/faith/2015/05/02/pope-hails-junipero-serra-one-founding-fathers-united-states.

81 **Pennsylvania passed its first fire ban:** In this section, the source of most of the quotations pertinent to fire policy in the United States is here: Omer Call Stewart, *Forgotten Fires: Native Americans and the Transient Wilderness* (Norman: University of Oklahoma Press, 2002).

81 **British officials lamented:** David Arnold, "Fire, Forest, City: A Social Ecology of Fire in British India," *Environment and History* 27, no. 3 (2021): 447–69, https://doi.org/10.3197/096734019X15463432086991.

81 **As French industrialists colonized:** Eugène Teston and Maurice Percheron, *L'Indochine Moderne; Encyclopédie Administrative, Touristique, Artistique et Économique* (Paris: Librarie de France, 1932).

81 **Even in Ireland:** Matthew S. Carroll, Catrin M. Edgeley, and Ciaran Nugent, "Traditional Use of Field Burning in Ireland: History, Culture and Contemporary Practice in the Uplands," *International Journal of Wildland Fire* 30, no. 6 (2021): 399–409, https://doi.org/10.1071/WF20127.

82 **the common people of Europe:** Stephen J. Pyne, *Vestal Fire: An Environmental History, Told through Fire, of Europe and Europe's Encounter with the World* (Seattle: University of Washington Press, 2012).

82 **sought out Linnaeus:** Most of the following information about Linnaeus, Baron Harleman, and fire comes from these sources: Michael R. Dove, "Linnaeus' Study of Swedish Swidden Cultivation: Pioneering Ethnographic Work on the 'Economy of Nature,'" *Ambio* 44, no. 3 (2015): 239–48; Stephen J. Pyne, "Maintaining Focus: An Introduction to Anthropogenic Fire," *Chemosphere* 29, no. 5 (1994): 889–911, https://doi.org/10.1016/0045-6535(94)90159-7; Helaine Selin,

ed., *Nature Across Cultures* (Dordrecht: Springer Netherlands, 2003). https://doi.org/10.1007/978-94-017-0149-5.

84 **criminalizing every tool:** See, for example, David Bollier, *Think Like a Commoner: A Short Introduction to the Life of the Commons* (Gabriola, BC: New Society Publishers, 2014).

84 **European governments during the Industrial Revolution:** Most of the quotes and some of the analysis in the latter part of this chapter come from Michael Perelman, *The Invention of Capitalism: Classical Political Economy and the Secret History of Primitive Accumulation* (Durham, NC: Duke University Press, 2000).

CHAPTER 5

94 **careers in wildlife:** Matthew Desmond, *On the Fireline: Living and Dying with Wildland Firefighters* (Chicago: University of Chicago Press, 2008).

100 **burrow into human flesh:** Ed Yong, "Fire Chasing Beetles Sense Infrared Radiation from Hundreds of Kilometers Away," *National Geographic*, May 27, 2012.

CHAPTER 6

105 **Fremont gave the order to attack:** The genocide that the U.S. federal government, the California state government, local governments, and citizen militias carried out against California's Indigenous people, while not widely discussed, is well documented. Much of the background on the origins and implementations of this genocide comes from Benjamin Madley's exhaustive *An American Genocide: The United States and the California Indian Catastrophe, 1846–1873* (New Haven, CT: Yale University Press, 2016). I also drew heavily from Brendan C. Lindsay's *Murder State: California's Native American Genocide, 1846–1873* (Lincoln: University of Nebraska Press, 2012). Kevin Starr offers another valuable treatment of the topic in *Rooted in Barbarous Soil: An Introduction to Gold Rush Society and Culture: People, Culture, and Community in Gold Rush California*, Kevin Starr and Richard J. Orsi, eds. (Berkeley: University of California Press, 2000). William J. Bauer Jr. provides Indigenous perspectives on the period in his books *California Through Native Eyes: Reclaiming History* (Seattle: University of Washington Press, 2016) and *We Are the Land: A History of Native California* (Berkeley: University of California Press, 2021).

107 **"It was a perfect butchery":** Kit Carson, *Kit Carson's Own Story of His Life* (Santa Fe: New Mexican Publishing Corp, 1955).

107 **"war of extermination":** Peter Burnett, "State of the State Address," delivered January 6, 1851, Governors Gallery, California State Library, https://governors.library.ca.gov/addresses/s_01-Burnett2.html.

107 **bill that criminalized the use of fire:** Kimberly Johnston-Dodds, "Early California Laws and Policies Related to California Indians," California State Library, California Research Bureau, September 2002, https://www.courts.ca.gov/documents/IB.pdf.

107 **"It became part of the policy":** This quote from Bill Tripp comes from Umair Irfan, "We Must Burn the West to Save It," *Vox*, October 22, 2020, https://www.vox.com/21507802/wildfire-2020-california-indigenous-native-american-indian-controlled-burn-fire.

108 **a general advised Congress:** Wade Davis, *The Wayfinders: Why Ancient Wisdom Matters in the Modern World* (Toronto, ON: House of Anansi Press, 2009).

109 **According to the *Daily Alta California*:** Benjamin Madley and Ben Kiernan, "'A War of Extermination': The California Indian Genocide, 1846–1873," in *The Cambridge World History of Genocide, Volume II: Genocide in the Indigenous, Early Modern and Imperial Worlds from c.1535 to World War One*, Ned Blackhawk, Ben Kiernan, Benjamin Madley, and Rebe Taylor, eds. (Cambridge: Cambridge University Press, 2023), 412–33.

111 **"the clearest case of genocide":** Robert V. Hine, John Mack Faragher, and Jon T. Coleman, *The American West: A New Interpretive History* (New Haven, CT: Yale University Press, 2017).

111 **reported two days prior:** Alex Wigglesworth, "Crews Tried but Couldn't Stop the Lava Fire Before It Became California's Worst of 2021 So Far," *Los Angeles Times*, July 3, 2021, www.latimes.com/california/story/2021-07-03/lava-fire-containment-fire-crews.

112 **As our hotshot buggies rumbled:** Sarah Kaplan, "Climate Change Has Gotten Deadly. It Will Get Worse," *Washington Post*, July 6, 2021, https://www.washingtonpost.com/climate-environment/2021/07/03/climate-change-heat-dome-death/.

115 **half its average snowpack:** "Water Year 2021: An Extreme Year," California Department of Water Resources, September 15, 2021, https://water.ca.gov/-/media/DWR-Website/Web-Pages/Water-Basics/Drought/Files/Publications-And-Reports/091521-Water-Year-2021-broch_v2.pdf.

115 **worst fires ever:** Akshay Kulkarni, "A Look Back at the 2021 B.C. Wildfire Season," CBC News, October 4, 2021, https://www.cbc.ca/news/canada/british-columbia/bc-wildfires-2021-timeline-1.6197751.

CHAPTER 7

129 **"Oh, this is bully!":** Most of the correspondence and conversations between Theodore Roosevelt and Gifford Pinchot in this chapter come from Timothy Egan, *The Big Burn: Teddy Roosevelt and the Fire That Saved America* (New York: Houghton Mifflin Harcourt, 2009).

129 **environmental historian William Cronon:** William Cronon, "The Trouble with Wilderness; Or, Getting Back to the Wrong Nature," *Environmental History* 1, no. 1 (1996): 7–28, https://doi.org/10.2307/3985059.
130 **"Never again," historian Frederick Jackson Turner:** Frederick Jackson Turner, *The Significance of the Frontier in American History* (Madison, WI: Silver Buckle Press, 1984).
130 **"The forest reserves":** Theodore Roosevelt, "State of the Union Addresses of Theodore Roosevelt," Gutenberg.org, https://www.gutenberg.org/files/5032/5032-h/5032-h.htm.
130 **a vile, godforsaken place:** Roderick Nash, *Wilderness and the American Mind* (New Haven, CT: Yale University Press, 1982).
132 **to make trees orderly:** James C. Scott, *Seeing Like a State: How Certain Schemes to Improve the Human Condition Have Failed* (New Haven, CT: Yale University Press, 2008).
133 **after burning Ahwahneechee villages:** Rebecca Solnit, *Savage Dreams: A Journey into the Landscape Wars of the American West* (Berkeley: University of California Press, 2014).
133 **While scholars have attributed:** Adriana Petryna, *Horizon Work: At the Edges of Knowledge in an Age of Runaway Climate Change* (Princeton, NJ: Princeton University Press, 2022).

CHAPTER 8

149 **William Greeley, a baby-faced man:** For information about the life of William Greeley, including his relationship with the logging industry and his enduring conflict with Gifford Pinchot, refer to his biography: George T. Morgan, *William B. Greeley: A Practical Forester, 1879–1955* (St. Paul, MN: Forest History Society, 1961), https://hdl.handle.net/2027/umn.319510001740795. For more contextual information, see Stephen J. Pyne, *Fire in America: A Cultural History of Wildland and Rural Fire* (Seattle: University of Washington Press, 1997).
149 **a "forest missionary":** Timothy Egan, *The Big Burn: Teddy Roosevelt and the Fire That Saved America* (New York: Houghton Mifflin Harcourt, 2009).
150 **"The public," he wrote in 1920:** William B. Greeley, "Paiute Forestry or the Fallacy of Light Burning," *Fire Management Today* 78, no. 1 (2020): 11–15.
150 **published in *The Timberman*:** Greeley, "Paiute Forestry or the Fallacy of Light Burning."
150 **"Absolute devastation," Pinchot wrote:** Egan, *The Big Burn*.
151 **West Coast Lumbermen's Association:** Harold K. Steen, *The U.S. Forest Service: A Centennial History* (Seattle: University of Washington Press, 2013).

151 **All "merchantable stumpage":** Greeley, "Paiute Forestry or the Fallacy of Light Burning."

157 **flashing with lightning:** Alex Wigglesworth, "California Wildfire Generates Its Own Lightning as It More Than Doubles in Size," *Los Angeles Times*, July 10, 2021, https://www.latimes.com/california/story/2021-07-10/wildfire-generates-its-own-lightning-as-it-more-than-doubles-in-size#:~:text=A%20wildfire%20in%20Northern%20California,conditions%20for%20firefighters%2C%20authorities%20said.

CHAPTER 9

161 **pilgrimage to Yarnell Hill:** The Yarnell Hill Fire is among the worst tragedies in the recent history of wildland firefighting. It has been the subject of numerous books that explore the incident from various angles. Brendan McDonough, the lone survivor, provides a personal account in his *My Lost Brothers: The Firsthand Account of a Tragic Wildfire, Its Lone Survivor, and the Firefighters Who Made the Ultimate Sacrifice* (New York: Hachette Books, 2016), later published as *Granite Mountain*, while Kyle Dickman's *On the Burning Edge* (New York: Ballantine Books, 2015) expands the scope to the lives of the men who died that day. Fernanda Santos's *The Fire Line* (New York: Flatiron Books, 2016) offers a comprehensive chronicle of the events leading up to the tragedy. Sean Flynn's article in *Outside* magazine, "19: The True Story of the Yarnell Hill Fire" (September 17, 2013, updated December 14, 2023), offers an excellent journalistic account of the fire.

166 **total wildland firefighter fatalities:** These statistics are captured and compiled by multiple agencies with varying inclusion and exclusion criteria, resulting in some differences in the data. For more information, see Corey Butler, Suzanne Marsh, Joseph W. Domitrovich, and Jim Helmkamp, "Wildland Firefighter Deaths in the United States: A Comparison of Existing Surveillance Systems," *Journal of Occupational and Environmental Hygiene* 14, no. 4 (2017): 258–70, https://doi.org/10.1080/15459624.2016.1250004; also see "NWCG Report on Wildland Firefighter Fatalities in the United States: 2007–2016," National Wildfire Coordinating Group, December 2017.

166 **The fire overtook fifteen firefighters:** Bill Gabbert, "15 Firefighters on Dolan Fire Become Entrapped by the Fire and Deployed Fire Shelters," Wildfire Today, September 8, 2020, https://wildfiretoday.com/2020/09/08/15-firefighters-on-dolan-fire-became-entrapped-by-the-fire-and-deployed-fire-shelters/.

167 **Were we really on our own:** The issue of health insurance for wildland firefighters is complicated and has been reported on extensively. James Puerini, a veteran wildland firefighter doing research at Yale, and Gerald Torres, also at Yale,

wrote about it here: James Puerini and Gerald Torres, "Don't Just Cheer Wildland Firefighters as Heroes. Give Them Affordable Healthcare," *Los Angeles Times*, June 20, 2020; also see Brianna Sacks, "Wildland Firefighters Are Relying on GoFundMe to Survive After Getting Injured on the Job," *BuzzFeed*, December 13, 2021.

167 **wife had been charged**: Sacks, "Wildland Firefighters Are Relying on GoFundMe to Survive After Getting Injured on the Job."

168 **file for bankruptcy:** Sacks, "Wildland Firefighters Are Relying on GoFundMe to Survive After Getting Injured on the Job."

168 **Since the early 1970s**: For reference, see Grassroots Wildland Firefighters, "(Mis)Classification," undated, https://www.grassrootswildlandfirefighters.com/misclassification.

168 **the institutional classification:** Patrick Withen, "Climate Change and Wildland Firefighter Health and Safety," *New Solutions* 24, no. 4 (2015): 577–84, https://doi.org/10.2190/NS.24.4.i.

169 **with knowledge of the consequences:** See, for example, Geoffrey Supran and Naomi Oreskes, "Assessing ExxonMobil's Climate Change Communications (1977–2014)," *Environmental Research Letters* 12, no. 8 (2017): 084019.

170 **Koch Industries is a case in point:** Much of the information about Koch Industries is drawn from Christopher Leonard, *Kochland: The Secret History of Koch Industries and Corporate Power in America* (London: Simon & Schuster, 2019).

170 **eliminate workers' rights:** For key details on this trend, see Nancy MacLean, *Democracy in Chains: The Deep History of the Radical Right's Stealth Plan for America* (New York: Viking, 2017).

171 **Grover Norquist, a Koch affiliate**: Benjamin Wallace-Wells, "The Long Afterlife of Libertarianism," *New Yorker*, May 29, 2023.

171 **California's Republican congressman Tom McClintock:** Bill Gabbert, "In Case You Missed the Congressman Calling Firefighters 'Unskilled Labor,'" *Wildfire Today*, July 25, 2021, https://wildfiretoday.com/2021/07/25/in-case-you-missed-the-congressman-calling-firefighters-unskilled-labor/.

172 **McClintock denies basic climate science:** See, for example, Bryan Anderson, "Trump, Climate Change, Carpetbagging: What You Missed from McClintock, Morse Debate," *Sacramento Bee*, updated September 24, 2018, https://www.sacbee.com/news/politics-government/capitol-alert/article218767065.html.

172 **signed a pledge to Charles Koch:** Jane Mayer, "Koch Pledge Tied to Congressional Climate Inaction," *New Yorker*, June 30, 2013, https://www.newyorker.com/news/news-desk/koch-pledge-tied-to-congressional-climate-inaction.

173 **Scientists call this a sacrifice zone:** Steve Lerner, *Sacrifice Zones: The Front Lines of Toxic Chemical Exposure in the United States* (Cambridge, MA: MIT Press, 2010).

173 **chronic health effects:** See, for example, Erin O. Semmens, Joseph Domitrovich, Kathrene Conway, and Curtis W. Noonan, "A Cross-Sectional Survey of Occupational History as a Wildland Firefighter and Health," *American Journal of Industrial Medicine* 59, no. 4 (2016): 330–35, https://doi.org/10.1002/ajim.22566. Also see Ana Isabel Miranda et al., "Wildland Smoke Exposure Values and Exhaled Breath Indicators in Firefighters," *Journal of Toxicology and Environmental Health, Part A* 75, nos. 13–15 (2012): 831–43, https://doi.org/10.1080/15287394.2012.690686.

173 **U.S. Department of Labor:** "Department of Labor Announces Streamlined Claims Process for Federal Firefighters with Certain Occupational Illnesses," U.S. Department of Labor, April 20, 2022, https://www.dol.gov/newsroom/releases/owcp/owcp20220420.

174 **surveyed wildland firefighters:** Grassroots Wildland Firefighters, "Partner/Spouse Survey 2021," https://www.grassrootswildlandfirefighters.com/partner spouse-survey.

175 **climate change was making everything worse:** Robin Cooper and Riva Duncan, "Special Report: Wildland Firefighters—Hidden Heroes of the Mental Health Effects of Climate Change," *Psychiatric News* 58, no. 5 (April 2023), https://doi.org/10.1176/appi.pn.2023.05.5.38.

176 **as many wildland firefighters die of suicide:** The data on suicide rates of wildland firefighters is variable, due in part to a culture that is generally averse to reporting on issues related to mental health. Ian H. Stanley, Melanie A. Hom, Anna R. Gai, and Thomas E. Joiner, "Wildland Firefighters and Suicide Risk: Examining the Role of Social Disconnectedness," *Psychiatry Research* 266 (August 2018): 269–74, https://doi.org/10.1016/j.psychres.2018.03.017.

CHAPTER 10

180 **billions of dollars in federal contracts:** Anna Fifield, "Contractors Reap $138bn from Iraq War," *Financial Times*, March 18, 2013, https://www.ft.com/content/7f435f04-8c05-11e2-b001-00144feabdc0.

181 **lightning fire in the Eastern Sierra:** Jim Carlton, "After Tamarack Fire, the U.S. Plans New Tactics to Fight West's Flames," *Wall Street Journal*, August 3, 2021, https://www.wsj.com/articles/tamarack-fire-california-nevada-11627939474.

187 **pacifist philosopher William James:** William James, *The Moral Equivalent of War, and Remarks at the Peace Banquet* (San Francisco: Center for Typographic Language, 2006). First published as Lecture 11 in *Memories and Studies* (New York: Longman Green and Co., 1911), 267–96.

187 **Harry Truman proposed:** "Program: The President's Conference on Fire Prevention, May 6–8, 1947, Departmental Auditorium, Washington, D.C.

Congressional Document," https://www.firehero.org/wp-content/uploads/2019/02/truman-report-1947.pdf.

190 **Fire camps are organized:** Elizabeth Weil, "They Know How to Prevent Megafires. Why Won't Anybody Listen?," *ProPublica*, August 28, 2020, https://www.propublica.org/article/they-know-how-to-prevent-megafires-why-wont-anybody-listen.

192 **the military's $778 billion budget:** "Defense Spending by State Fiscal Year 2020," U.S. Department of Defense, Office of Local Defense Community Cooperation, https://oldcc.gov/sites/default/files/defense-spending-rpts/OLDCC_DSBS_FY2020_FINAL_WEB.pdf.

192 **over half went to private corporations:** "Corporate Power, Profiteering, and the 'Camo Economy,'" Costs of War, Watson Institute for International and Public Affairs, Brown University, undated, https://watson.brown.edu/costsofwar/costs/social/corporate.

192 **cost taxpayers about $4.4 billion:** "National Large Incident Year-to-Date Report," National Interagency Fire Center, 2021, https://www.nifc.gov/fire-information/statistics/suppression-costs.

192 **150 firefighting companies:** Nick Mott, "Are Private Firefighters a Public Good or an Unfair Perk for the Wealthy?," KUNC, November 21, 2018, https://www.kunc.org/2018-11-21/are-private-firefighters-a-public-good-or-an-unfair-perk-for-the-wealthy.

192 **Forest Service awarded:** Sierra Dawn McClain, "Forest Service Awards $640M in Firefighting Contracts to Eight Oregon Firms," *Capital Press*, February 15, 2022, https://www.capitalpress.com/ag_sectors/timber/forest-service-awards-640m-in-firefighting-contracts-to-eight-oregon-firms/article_c4bf04fa-8eb9-11ec-9f42-33caa793f81b.html.

192 **Grayback Forestry, among America's largest:** Jamie Parfitt, "Local Forestry Firms Awarded US Forest Service Wildland Firefighting Contracts," NewsWatch 12, KDRV, February 10, 2022, https://www.kdrv.com/news/firewatch/local-forestry-firms-awarded-us-forest-service-wildland-firefighting-contracts/article_2e53195c-8aa4-11ec-8641-0f7c7e793b78.html.

193 **owned by Perimeter Solutions:** Samantha Masunaga, "It's a Staple of Fire Season, but What Is It?," *Los Angeles Times*, October 1, 2020, https://search.proquest.com/docview/2447463050?pq-origsite=primo.

193 **"more expensive than dropping Perrier":** Patty Nieberg, "Largest Firefighting Plane May Be Sold for COVID-19 Response," Oregon Public Broadcasting, April 28, 2021, https://www.opb.org/article/2021/04/28/largest-firefighting-plane-grounded-covid-19-possibilities/.

193 **a million dollars each day:** Brent McDonald, Sashwa Burrous, Eden Weingart, and Meg Felling, "Inside the Massive and Costly Fight to Contain the Dixie

Fire," *New York Times*, October 11, 2021, https://www.nytimes.com/interactive/2021/10/11/us/california-wildfires-dixie.html.

194 **Environmental regulations and reviews:** See Adriana Petryna, *Horizon Work: At the Edges of Knowledge in an Age of Runaway Climate Change* (Princeton, NJ: Princeton University Press, 2022). She deals with the topic of environmental regulations in chapter 7.

194 **If wildfires attract disaster capitalists:** For an excellent overview of this topic, see Naomi Klein, *The Shock Doctrine: The Rise of Disaster Capitalism* (New York: Metropolitan Books, 2007).

194 **billionaire Archie Aldis Emmerson:** Most of the information I use in this section of the chapter comes from the excellent reporting of Chloe Sorvina, "A Billion-Dollar Fortune from Timber and Fire," *Forbes*, May 14, 2018.

196 **president of Grayback Forestry:** Tony Schick, "How Fire Consumed the Forest Service Budget," Oregon Public Broadcasting, October 15, 2018, https://www.opb.org/news/article/wildfire-forest-service-budget-suppression-portion.

196 **denied the role of climate change:** Alix Langone, "Secretary Zinke Says Climate Change Is Not Responsible for California Wildfires, Blames Environmentalists," *Time*, August 13, 2018.

196 **accused of ethics violations:** Linda Qiu, "Ryan Zinke Broke Ethics Rules as Interior Secretary, Inquiry Finds," *New York Times*, February 16, 2022, https://www.nytimes.com/2022/02/16/us/politics/ryan-zinke-ethics.html.

196 **privatizing the public sector:** Greg Zimmerman, "Zinke Is Letting Corporations Profit Off Our National Parks," *High Country News*, August 31, 2018, http://www.hcn.org/articles/opinion-corporations-are-profiting-off-our-national-parks/.

196 **$100,000 per year lobbying Congress:** Schick, "How Fire Consumed the Forest Service Budget."

196 **Forest Service spent:** Associated Press, "Forest Service Testing New Wildfire Retardant, Critics Push Other Methods," CBS News Colorado, July 2, 2022, https://www.cbsnews.com/colorado/news/forest-service-testing-new-wildfire-retardant-critics-firefighters-environment/.

CHAPTER 11

201 **Global Environmental Crises:** Geoff Dembicki, *The Petroleum Papers: Inside the Far-Right Conspiracy to Cover Up Climate Change* (Vancouver, BC: Greystone Books, 2022).

202 **first informed the American Petroleum Institute:** Charles Jones, "A Review of the Air Pollution Research Program of the Smoke and Fumes Committee of the American Petroleum Institute," *Journal of the Air Pollution Control Association* 8, no. 3 (1958): 268–72, DOI: 10.108/00966665.1958.10467854.

202 **"Whenever you burn conventional fuel"**: Allan Nevins and Courtney C. Brown, *Energy and Man: A Symposium* (New York: Muriwai Books, 2018).
202 **"There is still time"**: Benjamin Franta, "Early Oil Industry Knowledge of CO_2 and Global Warming," *Nature Climate Change* 8, no. 12 (2018): 1024–25, https://doi.org/10.1038/s41558-018-0349-9.
202 **Stanford physicist John Laurman warned:** "CO2 and Climate Task Force (AQ-9), Minutes of Meeting," 1980, https://www.documentcloud.org/documents/3483045-AQ-9-Task-Force-Meeting-1980.html.
202 **memo circulated among Exxon managers:** Roger W. Cohen, "1981 Exxon Memo on Possible Emission Consequences of Fossil Fuel Consumption," Climatefiles, https://www.climatefiles.com/exxonmobil/1981-exxon-memo-on-possible-emission-consequences-of-fossil-fuel-consumption/.
202 **Exxon report predicted with precision:** M. B. Glaser, "1982 Memo to Exxon Management about CO2 Greenhouse Effect," Climatefiles, https://www.climatefiles.com/exxonmobil/1982-memo-to-exxon-management-about-co2-greenhouse-effect/.
202 **Royal Dutch Shell's scientists:** Greenhouse Effect Working Group, "1988 Shell Confidential Report: 'The Greenhouse Effect,'" May 28, 1988, Climatefiles, https://www.climatefiles.com/shell/1988-shell-report-greenhouse/.
202 **said Lew Ward:** Dembicki, *The Petroleum Papers*.
204 **California's temperatures have risen:** "Climate Change in California," U.S. Environmental Protection Agency, September 2016, https://19january2017snapshot.epa.gov/sites/production/files/2016-09/documents/climate-change-ca.pdf.
204 **major changes in plant flammability:** See, for example, Rachael H. Nolan et al., "Predicting Dead Fine Fuel Moisture at Regional Scales Using Vapour Pressure Deficit from MODIS and Gridded Weather Data," *Remote Sensing of Environment* 174 (March 2016): 100–108, https://doi.org/10.1016/j.rse.2015.12.010. Also see Alexandra D. Syphard, Jon E. Keeley, and T. J. Brennan, "High-Resolution Wildfire Simulations Reveal Complexity of Climate Change Impacts on Projected Burn Probability for Southern California," *Fire Ecology* 17, no. 1 (2021): 1–20, DOI: 10.1186/s42408-021-00106-9.
204 **more flammable for more of the year:** Michael Goss et al., "Climate Change Is Increasing the Likelihood of Extreme Autumn Wildfire Conditions Across California," *Environmental Research Letters* 15, no. 9 (2020): 094016, https://doi.org/10.1088/1748-9326/ab83a7.
204 **extended fire seasons:** "USDA Forest Service and Partners Gear Up for Significant 2016 Wildfire Season," U.S. Department of Agriculture, May 17, 2016, https://www.usda.gov/media/press-releases/2016/05/17/usda-forest-service-and-partners-gear-significant-2016-wildfire.
204 **area facing severe fire risk:** Hayley Smith and Sean Greene, "More Than Half

of Rural California Now Ranks 'Very High' for Wildfire Hazard," *Los Angeles Times*, December 27, 2022, https://search.proquest.com/docview/2758222669?pq-origsite=primo.

204 **now the most significant driver:** See Yizhou Zhuang et al., "Quantifying Contributions of Natural Variability and Anthropogenic Forcings on Increased Fire Weather Risk over the Western United States," *Proceedings of the National Academy of Sciences* 118, no. 45 (2021): e2111875118, https://doi.org/10.1073/pnas.2111875118.

206 **accepted across party lines:** Elizabeth Kolbert, "How Did Fighting Climate Change Become a Partisan Issue?," *New Yorker*, August 14, 2022, https://www.newyorker.com/magazine/2022/08/22/how-did-fighting-climate-change-become-a-partisan-issue.

206 **prominent pesticide was decimating ecosystems:** The information in this paragraph comes from Naomi Oreskes and Erik M. Conway, *Merchants of Doubt: How a Handful of Scientists Obscured the Truth on Issues from Tobacco Smoke to Global Warming* (New York: Bloomsbury Press, 2010).

207 **Charles and David Koch:** Christopher Leonard, *Kochland: The Secret History of Koch Industries and Corporate Power in America* (London: Simon & Schuster, 2019).

207 **according to biographer Christopher Leonard:** Leonard, *Kochland*.

207 **Harvard scholar Naomi Oreskes:** Oreskes and Conway, *The Big Myth*.

207 **strategies were already well established:** Michael E. Mann, *The New Climate War: The Fight to Take Back Our Planet* (Brunswick, Victoria, Australia: Scribe Publications, 2021).

207 **Their proxy groups:** See Nancy MacLean, *Democracy in Chains: The Deep History of the Radical Right's Stealth Plan for America* (New York: Viking, 2017).

207 **Doubt was their product:** Oreskes and Conway, *The Big Myth*.

208 **the Koch team deployed:** Theda Skocpol and Alexander Hertel-Fernandez, "The Koch Network and Republican Party Extremism," *Perspectives on Politics* 14, no. 3 (2016): 681–99, https://doi.org/10.1017/S1537592716001122.

208 **fund election challenges:** See, for example, Leah Cardamore Stokes, *Short Circuiting Policy: Interest Groups and the Battle over Clean Energy and Climate Policy in the American States* (New York: Oxford University Press, 2020).

208 **fossil fuel industry's campaign donations:** Open Secrets, "Oil & Gas Summary," 2016, OpenSecrets, https://www.opensecrets.org/industries//totals?cycle=2016&ind=E01.

208 **8 of 278 Republicans:** Julie Kliegman, "Jerry Brown Says 'Virtually No Republican' in Washington Accepts Climate Change Science," Politifact, May 18, 2014, https://www.politifact.com/factchecks/2014/may/18/jerry-brown/jerry-brown-says-virtually-no-republican-believes-/.

208 **more fossil fuels:** Throfinn Stainforth and Bartosz Brzezinski, "More Than Half of All CO2 Emissions Since 1751 Emitted in the Last 30 Years," Institute

for European Environmental Policy, April 29, 2020; see also Bill McKibben, "A Very Hot Year," *New York Review of Books*, March 12, 2020, https://www.nybooks.com/articles/2020/03/12/climate-change-very-hot-year/.

208 **$3.2 billion in profits:** Damian Carrington, "Revealed: Oil Sector's 'Staggering' $3bn-a-Day Profits for Last 50 Years," *Guardian*, July 21, 2022, https://www.theguardian.com/environment/2022/jul/21/revealed-oil-sectors-staggering-profits-last-50-years.

210 **paramedics ran out of cooling stations:** Dhruv Khullar, "What a Heat Wave Does to Your Body," *New Yorker*, August 25, 2023, https://www.newyorker.com/news/annals-of-a-warming-planet/what-a-heat-wave-does-to-your-body.

210 **cheapest and most reliable:** See "Massive Expansion of Renewable Power Opens Door to Achieving Global Tripling Goal Set at COP28," International Energy Agency, January 11, 2024, https://www.iea.org/news/massive-expansion-of-renewable-power-opens-door-to-achieving-global-tripling-goal-set-at-cop28; see also "Renewable Power Generation Costs in 2021," International Renewable Energy Agency, July 13, 2022, https://www.irena.org/publications/2022/Jul/Renewable-Power-Generation-Costs-in-2021.

210 **five pathways to net-zero:** "Net-Zero America: Potential Pathways, Infrastructure, and Impacts," Princeton University, 2021, https://netzeroamerica.princeton.edu/?explorer=year&state=national&table=2020&limit=200.

211 **61 percent finally accepted:** This statistic is drawn from a Pew poll and tracks the percentage of Republicans who believe that human activity contributes a great deal (14 percent) or some (39 percent) to climate change, which for the purposes of this chapter counts as being more or less aligned with the scientific consensus. See Cary Funk and Meg Hefferon, "U.S. Public Views on Climate and Energy," Pew Research Center Science & Society, November 25, 2019, https://www.pewresearch.org/science/2019/11/25/u-s-public-views-on-climate-and-energy/.

211 **Ben Shapiro tweeted:** Ben Shapiro, Twitter post, September 14, 2020, 2:00 PM, https://twitter.com/benshapiro/status/1305569639653498880.

211 **smoke from California's fires reached Europe:** Holger Baars et al., "Californian Wildfire Smoke over Europe: A First Example of the Aerosol Observing Capabilities of Aeolus Compared to Ground-Based Lidar," *Geophysical Research Letters* 48, no. 8 (2021): e2020GL092194, https://doi.org/10.1029/2020GL092194.

211 **$200 million lobbying:** See, for example, Sandra Laville, "Top Oil Firms Spending Millions Lobbying to Block Climate Change Policies, Says Report," *Guardian*, March 22, 2019, https://www.theguardian.com/business/2019/mar/22/top-oil-firms-spending-millions-lobbying-to-block-climate-change-policies-says-report.

211 **industry front groups:** Bill McKibben, "From Climate Exhortation to Climate Execution," *New Yorker*, December 27, 2022, https://www.newyorker.com/news/daily-comment/from-climate-exhortation-to-climate-execution.

212 **"flood the zone with shit":** Paul Starr, "The Flooded Zone: How We Became More Vulnerable to Disinformation in the Digital Era," in *The Disinformation Age*, Steven Livingston and W. Lance Bennett, eds. (New York: Cambridge University Press, 2020), 67–92, https://doi.org/10.1017/9781108914628.003.

212 **Fossil fuel pundits:** Christina Larson, Jennifer McDermott, Patrick Whittle, and Wayne Perry, "Contrary to Politicians' Claims, Offshore Wind Farms Don't Kill Whales. Here's What to Know," AP News, December 23, 2023, https://apnews.com/article/offshore-wind-whales-deaths-trump-5158af7f5bf0f5ef9e1530564ff791a9.

212 **Tennessee officially labeled methane gas:** "HB0946—Tennessee 113th General Assembly (2023–2024)," Tennessee House Bill 946, passed April 18, 2023, LegiScan, https://legiscan.com/TN/text/HB0946/id/2784209.

212 **describes as "unproven":** "Technical Summary, Global Warming of 1.5°C: An IPCC Special Report on the Impacts of Global Warming of 1.5°C Above Pre-Industrial Levels and Related Global Greenhouse Gas Emission Pathways," Intergovernmental Panel on Climate Change, 2018, https://www.ipcc.ch/site/assets/uploads/sites/2/2018/12/SR15_TS_High_Res.pdf.

212 **Jewish space lasers:** As *Vox*'s Beauchamp points out, Greene said the space lasers were owned by the Rothschild family, which has long been a not-so-subtle means of disparaging Jewish people in general. You can read his analysis here: Zack Beauchamp, "Marjorie Taylor Greene's Space Laser and the Age-Old Problem of Blaming the Jews," *Vox*, January 30, 2021, https://www.vox.com/22256258/marjorie-taylor-greene-jewish-space-laser-anti-semitism-conspiracy-theories.

212 **right-wing militias in Oregon:** See Luke Mogelson, *The Storm Is Here: An American Crucible* (New York: Penguin Press, 2022).

213 **Our current trajectory:** Laura Pulido, "Racism and the Anthropocene," in *Future Remains*, Gregg Mitman, Marco Armiero, and Robert Emmett, eds. (Chicago: University of Chicago Press, 2018), 116–28, https://doi.org/10.7208/9780226508825-014.

CHAPTER 12

221 **The giant sequoia:** There are many excellent books on the sequoias that detail both their natural and cultural history. In this chapter, I drew heavily on Lori Vermass's *Sequoia: The Heralded Tree in American Art and Culture* (Washington, D.C.: Smithsonian Books, 2003) and William Tweed's *King Sequoia: The Tree That Inspired a Nation, Created Our National Park System, and Changed the Way We Think about Nature* (Berkeley, CA: Heyday Books, 2016). I also referred to *Ancient Sentinels: The Sequoias of Yosemite National Park* by Thomas Harvey, Bill Kuhn, and Pete Devine (San Francisco: Yosemite Conservancy, 2013); *The Giant*

Sequoia of the Sierra Nevada by Richard Hartesveldt (Washington, D.C.: U.S. Department of the Interior, 1975); and *The Once and Future Forest: California's Iconic Redwoods* by Sam Hodder et al. (Berkeley, CA: Heyday Books, 2021).

227 **defend their territory:** For more details, see Gelya Frank, *Defying the Odds: The Tule River Tribe's Struggle for Sovereignty in Three Centuries* (New Haven, CT: Yale University Press, 2010).

227 **bulldozing their sacred sites:** For a critique of U.S. Forest Service tribal consultation during the Windy Fire, see Jeanine Pfeiffer, "Wildfire Management and Recovery on Tribal Lands Complicated by Policy Inequities," PBS SoCal, August 18, 2022, https://www.pbssocal.org/news-community/wildfire-management-and-recovery-on-tribal-lands-complicated-by-policy-inequities.

231 **10 to 14 percent:** Nathaniel Stephenson and Christy Brigham, "Preliminary Estimates of Sequoia Mortality in the 2020 Castle Fire," National Park Service, June 25, 2021, https://www.nps.gov/articles/000/preliminary-estimates-of-sequoia-mortality-in-the-2020-castle-fire.htm.

233 **currently more giant sequoias alive:** Thanks to Sam Boland for bringing this tidbit to my attention. Rebecca Morelle and Alison Francis, "Giant Redwoods: World's Largest Trees 'Thriving in UK,'" BBC, March 12, 2024, https://www.bbc.com/news/science-environment-68518623.

234 *National Geographic* **published:** David Quammen, "Forest Giant," *National Geographic*, December 1, 2012.

234 **among the most fire-evolved:** Nancy Muleady Mecham, Catherine Wilcoxson Ueckert, and Ivo Lindauer, "Fire & the Natural History of Giant Sequoias," *American Biology Teacher* 64, no. 8 (2002): 573–78, https://doi.org/10.1662/0002-7685%282002%29064%5B0573%3AFTNHOG%5D2.0.CO%3B2.

234 **sequoias began to die:** Craig Welch, "Forests Are Reeling from Climate Change—but the Future Isn't Lost," *National Geographic*, April 14, 2022, https://www.nationalgeographic.com/magazine/article/forests-future-threatened-heat-drought-feature.

243 **"We are racing against the clock":** Charles Whisand, "Sequoias Amendment Included in Funding Package," *Porterville Recorder*, July 21, 2022, https://www.recorderonline.com/news/sequoias-amendment-included-in-funding-package/article_57f7b312-094d-11ed-bab3-f7884838571b.html.

243 **forty thousand new oil wells:** Veronica Morley, "Planning Commission Recommends Adding New Oil Wells," 23ABC News Bakersfield, CA, February 12, 2021, https://www.turnto23.com/news/local-news/planning-commission-recommends-adding-new-oil-wells.

244 **In classic doublespeak:** John Cox, "Newsom Orders End to Fracking in California by 2024, Work Toward Phasing Out All In-State Oil Production by 2045,"

Bakersfield Californian, April 23, 2021, https://www.bakersfield.com/news/newsom-orders-end-to-fracking-in-california-by-2024-work-toward-phasing-out-all-in/article_8200b8ba-a464-11eb-be8b-4b83ebb0e6a3.html.

244 **Kern County already accounts:** James Brock, "CRC to Halt Most Drilling if Ruling Stands," *Los Angeles Business Journal*, March 18, 2024, https://labusinessjournal.com/featured/crc/.

244 **Fong prefers forest management:** Vince Fong, "Wildfires," Vince Fong for Congress, https://vincefong.com/issues/wildfires/.

CHAPTER 13

249 **Climate-driven disasters:** Adam B. Smith, "2021 U.S. Billion-Dollar Weather and Climate Disasters in Historical Context," NOAA, Climate.gov, January 24, 2022, https://www.climate.gov/news-features/blogs/beyond-data/2021-us-billion-dollar-weather-and-climate-disasters historical#:~:text=In%202021%2C%20the%20U.S.%20experienced,billion%2Ddollar%20events%20in%202020.

249 **extinction of multiple species:** "23 Species from 19 States Lost to Extinction," Center for Biological Diversity, September 29, 2021, https://biologicaldiversity.org/w/news/press-releases/23-species-from-19-states-lost-to-extinction-2021-09-29/.

CHAPTER 14

269 **the memorandum read:** "Agreement for Shared Stewardship of California's Forest and Rangelands between the State of California and the USDA, Forest Service Pacific Southwest Region," California Governor's Office, August 12, 2020, https://www.gov.ca.gov/wp-content/uploads/2020/08/8.12.20-CA-Shared-Stewardship-MOU.pdf.

273 **the Forest Watch claimed:** "Over 99.9% of Public Comments Oppose Pine Mountain Project According to Analysis," Los Padres Forest Watch, April 6, 2021, https://lpfw.org/over-99-9-of-public-comments-oppose-pine-mountain-project-according-to-analysis/.

273 **four times more fuel:** There are many ways of calculating this. Malcolm North, in an interview with me, thought that this was a safe estimate.

275 **California forests are not "natural":** R. K. Hagmann et al., "Evidence for Widespread Changes in the Structure, Composition, and Fire Regimes of Western North American Forests," *Ecological Applications* 31, no. 8 (2021): e02431, https://doi.org/10.1002/eap.2431.

275 **ten thousand years of Indigenous management:** See Deniss J. Martinez et al.,

"Indigenous Fire Futures: Anticolonial Approaches to Shifting Fire Relations in California," *Environment and Society* 14, no. 1 (2023): 142–61, https://doi.org/10.3167/ares.2023.140109.

277 **I knew that chaparral thrives:** Caitlin L. Lippitt, Douglas A. Stow, John F. O'Leary, and Janet Franklin, "Influence of Short-Interval Fire Occurrence on Post-Fire Recovery of Fire-Prone Shrublands in California, USA," *International Journal of Wildland Fire* 22, no. 2 (2013): 184–93, https://doi.org/10.1071/WF10099.

277 **effectiveness of fuels treatments:** Scott Stephens, a professor in Berkeley's Department of Environmental Science, Policy, and Management, is an authority on this topic. You can view a summary of his research here: "Sierra Nevada Forest Restoration Works, https://forests.berkeley.edu/sites/forests.berkeley.edu/files/The%20Fire%20and%20Fire%20Surrogate%20Study%20Summary_2.pdf. For more recent research, see Scott L. Stephens et al., "Fire and Climate Change: Conserving Seasonally Dry Forests Is Still Possible," *Frontiers in Ecology and the Environment* 18, no. 6 (2020): 354–60, https://doi.org/10.1002/fee.2218; Shawn T. McKinney, Ilana Abrahamson, Theresa Jain, and Nathaniel Anderson, "A Systematic Review of Empirical Evidence for Landscape-Level Fuel Treatment Effectiveness," *Fire Ecology* 18, no. 1 (2022): 21, https://doi.org/10.1186/s42408-022-00146-3.

279 **preponderance of data:** For a small sample, see Susan J. Prichard, Nicholas A. Povak, Maureen C. Kennedy, and David W. Peterson, "Fuel Treatment Effectiveness in the Context of Landform, Vegetation, and Large, Wind-Driven Wildfires," *Ecological Applications* 30, no. 5 (2020): e02104, https://doi.org/10.1002/eap.2104; Susan J. Prichard et al., "Adapting Western North American Forests to Climate Change and Wildfires: 10 Common Questions," *Ecological Applications* 31, no. 8 (2021): e02433, https://doi.org/10.1002/eap.2433; S. L. Stephens et al., "Managing Forests and Fire in Changing Climates," *Science* 342, no. 6154 (2013): 41–42, https://doi.org/10.1126/science.1240294; S. L. Stephens et al., "Forest Restoration and Fuels Reduction: Convergent or Divergent?," *BioScience* 71, no. 1 (2021): 85–101, https://doi.org/10.1093/biosci/biaa134; Douglas J. Tempel et al., "Evaluating Short- and Long-Term Impacts of Fuels Treatments and Simulated Wildfire on an Old-Forest Species," *Ecosphere* 6, no. 12 (2015): 1–18, https://doi.org/10.1890/ES15-00234.1.

279 **Jordan Fisher Smith:** Jordan Fisher Smith, "Fire Wars," *OpenMind*, September 22, 2022, https://www.openmindmag.org/articles/fire-wars.

283 **opposition on false claims:** For an overview of misinformation regarding wildfires, prescribed burns, and fuels treatments, see Gavin M. Jones et al., "Counteracting Wildfire Misinformation," *Frontiers in Ecology and the Environment* 20, no. 7 (2022): 392–93, https://doi.org/10.1002/fee.2553; M. Zachariah Peery

et al., "The Conundrum of Agenda-Driven Science in Conservation," *Frontiers in Ecology and the Environment* 17, no. 2 (2019): 80–82, https://doi.org/10.1002/fee.2006.

CHAPTER 15

286 **expanding burn operations:** Rebecca K. Miller, Christopher B. Field, and Katharine J. Mach, "Barriers and Enablers for Prescribed Burns for Wildfire Management in California," *Nature Sustainability* 3, no. 2 (2020): 101–9, https://doi.org/10.1038/s41893-019-0451-7.

286 **reduced by around 95 percent:** Sam Hodder et al., *The Once and Future Forest: California's Iconic Redwoods* (Berkeley, CA: Heyday Books, 2018).

291 **largest in state history:** Susan Montoya Bryan, "Miscalculations, Errors Blamed for Massive New Mexico Blaze," *Los Angeles Times*, June 21, 2022.

291 **arrested a Forest Service burn boss:** Gabrielle Canon, "The Arrest That Shocked the Firefighting World—and Threatens a Vital Practice," *Guardian*, November 13, 2022, https://www.theguardian.com/world/2022/nov/13/prescribed-burn-fire-chief-arrest-oregon-new-mexico.

292 **now bloom weeks earlier:** For more information about this general topic, see David W. Inouye, "Climate Change and Phenology," *Wiley Interdisciplinary Reviews: Climate Change* 13, no. 3 (2022): e764-n/a, https://doi.org/10.1002/wcc.764.

292 **the carbon emitted:** See Michael Jerrett, Amir S. Jina, and Miriam E. Marlier, "Up in Smoke: California's Greenhouse Gas Reductions Could Be Wiped Out by 2020 Wildfires," *Environmental Pollution* 310 (2022): 119888, https://doi.org/10.1016/j.envpol.2022.119888.

292 **triple the annual footprint:** Leyland Cecco, "Wildfires Turn Canada's Vast Forests from Carbon Sink into Super-Emitter," *Guardian*, September 22, 2023, https://www.theguardian.com/world/2023/sep/22/canada-wildfires-forests-carbon-emissions.

292 **transform the Amazon rainforest:** Chelsea Harvey, "Wildfires Could Transform Amazon from Carbon Sink to Source," *Scientific American*, January 14, 2020, https://www.scientificamerican.com/article/wildfires-could-transform-amazon-from-carbon-sink-to-source/.

292 **net contributors to climate change:** Minho Kim, "Forests Are Losing Their Ability to Hold Carbon," *Scientific American*, July 26, 2023, https://www.scientificamerican.com/article/forests-are-losing-their-ability-to-hold-carbon/.

292 **"stop burning things":** Toward the end of this article, McKibben does acknowledge the importance of landscape burning for Indigenous people. Bill McKibben, "In a World on Fire, Stop Burning Things," *New Yorker*, March 18, 2022.

292 **burning landscapes the right way:** Adam Pellegrini et al., "Fire Effects on the Persistence of Soil Organic Matter and Long-Term Carbon Storage," *Nature Geoscience* 15, no. 1 (2022): 5–13, https://doi.org/10.1038/s41561-021-00867-1.

293 **thought this ban was wrongheaded:** Alex Wigglesworth, "Forest Service Resumes Prescribed Fire Program, but Some Fear New Rules Will Delay Projects," *Los Angeles Times*, September 13, 2022.

293 **success rate of 99.84 percent:** Randy Moore, "National Prescribed Fire Program Review Released (video)," United States Forest Service, September 8, 2022, https://www.fs.usda.gov/inside-fs/leadership/national-prescribed-fire-program-review-released-video.

294 **fleeing the Forest Service:** Jeva Lange, "Everyone Agrees Wildland Firefighters Deserve a Raise. Why Can't Congress Make It Happen?" *Heatmap News*, February 27, 2024, https://heatmap.news/politics/wildland-firefighter-salary-congress.

CHAPTER 16

306 **beginning to collaborate:** Tony Marks-Block and William Tripp, "Facilitating Prescribed Fire in Northern California through Indigenous Governance and Interagency Partnerships," *Fire* 4, no. 3 (2021): 37, https://doi.org/10.3390/fire4030037.

307 **said Robyne Redwater:** Harrison Tasoff, "A Chumash Cultural Burn Reignites Ancient Practice for Wilderness Conservation," *Santa Barbara Independent*, October 7, 2023, https://www.independent.com/2023/10/07/a-chumash-cultural-burn-reignites-ancient-practice-for-wilderness-conservation/.

307 **leveraged to reinforce:** See, for example, Morning Star Gali, "Stolen Freedom: The Ongoing Incarceration of California's Indigenous Peoples," in *Abolition for the People*, C. Kaepernick, ed. (New York: Kaepernick Publishing, 2021), 67–72.

307 **Indigenous burn practices:** Deniss J. Martinez et al., "Indigenous Fire Futures: Anticolonial Approaches to Shifting Fire Relations in California," *Environment and Society* 14, no. ss1 (2023): 142–61, https://doi.org/10.3167/ares.2023.140109.